Demographic Research Monographs

A Series of the Max Planck Institute
for Demographic Research, Rostock, Germany

Editor-in-chief: James W. Vaupel

Demographic Research Monographs

E. Barbi, S. Bertino, E. Sonnino (Eds.)
Inverse Projection Techniques
2004. XI, 137 pages. 3-540-20931-X

Gabriele Doblhammer

The Late Life Legacy of Very Early Life

With 38 Figures and 23 Tables

Springer

PD Dr. Gabriele Doblhammer
Max Planck Institute for Demographic Research
Konrad-Zuse-Straße 1
18057 Rostock
Germany
doblhammer@demogr.mpg.de

Printed with the financial support of the Max Planck Society

ISBN 3-540-22105-0 Springer Berlin Heidelberg New York

Cataloging-in-Publication Data:
Library of Congress Control Number: 2004107863

Springer is a part of Springer Science+Business Media

springeronline.com

Printed in Germany

Hardcover-Design: Erich Kirchner, Heidelberg
The front cover photo – top right-hand corner – includes a photograph by Harald Wenzel-Orf (www.wenzel-orf.de) and has been reproduced with his permission.

SPIN 11011453 42/3130-5 4 3 2 1 0 – Printed on acid-free paper

Foreword

This stimulating, carefully-researched book on *The Late Life Legacy of Very Early Life* by Dr. Gabriele Doblhammer is the second volume of a new series of Demographic Research Monographs published by Springer Verlag. The topic of the book is fascinating. Is a person's lifespan influenced by health and nutrition *in-utero* and shortly after birth? If so, why? The answers uncovered by the diligence, demographic and statistical expertise, and probing intelligence of the author are surprising but convincing.

To pry open the mystery of the lingering impact of very early life, Dr. Doblhammer focuses on month of birth. It turns out that people born in some months live substantially longer on average than people born in other months, not because of astrological forces but for reasons of health and nutrition. Dr. Doblhammer was educated in statistics and demography and earlier this year was the first person ever to receive the "Habilitation" degree, the recognition given in the German-speaking world to proven scholars who are qualified to become professors, in Demography. This book, which is evidence that she fully deserves this award, will not only provide important new findings about the legacy of early life but will also serve as a comprehensive foundation of knowledge on which future scholars can build.

The Demographic Research Monographs series can be considered the successor to the series called Odense Monographs on Population Aging, edited by Bernard Jeune and James Vaupel. The volumes in this now-terminated series were first published as hardcover books by an academic publisher, the Odense University Press, and subsequently made available online at www.demogr.mpg.de/books/odense. The nine Odense Monographs on Population Aging include two collections of research articles that focus on specific subjects on the frontier of demographic research, three volumes by senior researchers that present path-breaking findings, a review of research on a topic of emerging interest, a presentation of a new method for analysis of demographic data, an out-standing doctoral dissertation, and a unique collection of important demo-graphic data on non-human species.

The new series of Demographic Research Monographs will continue this mix, with books that are often under 200 pages in length, that have a clear focus, and that significantly advance demographic knowledge. Research related to population aging will continue to be a prime focus on the new series, but not the only one. The new series will embrace all of demography, broadly defined. As indicated by the first volume, an important subject will be historical demography. We also plan to highlight research on fertility and family dynamics, especially in Europe. Mathematical demography is the core of the population sciences and we will strive to foster monographs that use mathematics and statistics to further develop the theories and methods of demography. Biodemography is a small but rapidly growing and particularly innovative branch of demography: we will seize opportunities to publish monographs at the intersection of biology and demography, pertaining both to human and to other species, and including demographic research with ties to such fields as epidemiology, genetics, evolutionary biology, life-history biology, experimental demography, and paleodemography.

Each volume in the Demographic Research Monograph series will have a substantial link to the Max Planck Institute for Demographic Research. This volume *The Late Life Legacy of Very Early Life*, is written by Dr. Gabriele Doblhammer, who is a senior research scientist at the Institute. The third volume, *Supercentenarians*, is the outgrowth of four research workshops organized by the Max Planck Institute. And the fourth volume, tentatively entitled *To Long Life!?*, by Prof. Bernard Jeune, is a revised and updated version of a book that was originally published in Danish: the Max Planck Institute is supporting the additional work for this book.

As well as being published as hardcover books by Springer-Verlag, this and the following volumes of the Max Planck series of Demographic Research Monographs will subsequently be available at www.demogr.

mpg.de/books/drm. The online version may include color graphs, supplemental analyses, databases and other ancillary or enhanced material. Parallel publication online and in print is a significant innovation that will make the monograph series particularly useful to scholars and students around the world.

James W. Vaupel
Editor-in-Chief

Preface

In the 1970s Forsdahl, a medical doctor in Norway, observed that regional differences in adult lung cancer and heart disease were not related to contemporary differences in lifestyle, smoking behavior, or socioeconomic status but rather to differences in regional infant mortality during childhood and youth of the cohorts under study. His study is now considered to be the starting point of a large and productive area of research that tries to link early-life conditions to the manifestation of chronic disease later in life. The discussion about early-life effects on health at adult ages gained momentum with studies conducted by the Southampton group of Barker and colleagues. The group developed the fetal-origins hypothesis of adult disease (also known as the 'Barker hypothesis'), which suggests that coronary heart disease at adult ages results from poor conditions *in-utero* caused by inadequate nutrition on the part of the mother and infectious diseases she suffered during pregnancy. Since inadequate nutrition of the fetus is reflected in low birth weight, the Barker hypothesis claims that growth retardation *in-utero* leads to low birth weight and to an increased risk of chronic disease later in life.

Critics of the hypothesis frequently bring forward the argument that birth weight is confounded with socioeconomic status. Negative social factors in the early-life environment may set people onto life trajectories that negatively affect their health over the whole life course. Therefore, the almost universally observed relationship between birth weight and the risk of chronic disease later in life may be an outcome of the whole life course rather than the result of a critical period early in life.

This criticism leads to the question whether one can find an indicator for the prenatal and early postnatal environment that is not related to the life-course. This monograph shows that month of birth can be used as an indicator for the seasonally changing environment around the time of birth. It tests a series of alternative hypotheses which all assume the existence of life-course factors or social factors that are related to the month of birth. Only after having rejected all these hypotheses is a thorough analysis of the possible causal mechanisms performed.

I would like to thank James W. Vaupel, Founding Director of the Max Planck Institute for Demographic Research, for all his support and encouragement that he has given to me in the preparation of this monograph. I thank Kaare Christensen, Phil Cook, Jutta Gampe, Jan Hoem and Wolfgang Lutz for their valuable comments and Josef Kytir and Axel Skythe for help with data collection. I also express my gratitude to Karl Brehmer and Renee Flibotte for English language editing; and Elena Muth and Matthias Zenker for help with layout and graphic design.

Gabriele Doblhammer

Table of Contents

1 The Demographic and Epidemiological Context

When is a person's lifespan determined? At conception through genes, *in-utero*, during childhood and adolescence, or at adult ages through environmental factors? Different opinions prevailed at different times in the course of the 20th century and are still hotly debated. Infancy and childhood were seen as the critical phase in the first thirty years of the twentieth century. Since then the emphasis has been more on lifestyle factors and, thus, adult ages. Since the 1970s, however, new evidence has been emerging that chronic disease is rooted in infancy and childhood. An influential theory developed in the course of the last decade traces the causes of heart disease and diabetes back to the womb. The theory claims that the mother's nutrition and any infectious diseases she might have had during pregnancy are responsible for an increased susceptibility of the child to heart disease and diabetes once it reaches adult ages. The interpretation of this recent research has been challenged, however. Deprivation *in-utero* and during childhood may simply be the starting point of deprivation throughout life. The increased susceptibility later in life may thus be the result of the life-long accumulation of detrimental effects.

The month of birth fits into the debate about whether chronic disease at adult ages is determined by "programming" *in-utero* or in childhood on the one hand, or whether early-life factors are mediated through the lifelong accumulation of detrimental effects. To contribute to the debate, an indicator of the environment early in life is needed which does not have any life course interpretation. Starting with the work of Ellsworth Huntington (1934) a large body of epidemiological research (see Chapter 5) proposes that month of birth is just such an indicator. This monograph takes up this idea and explores the month-of-birth pattern in survival and causes of death. An extensive part of this monograph, however, is devoted to the question whether month of birth is indeed free of social confounding and of life-course interpretation.

If month of birth is a measure of the seasonal changes in the environment early in life and is independent of life-course factors, then the differences by month of birth indicate that the very first period of life influences late-life mortality and the susceptibility to chronic disease.

This chapter follows the debate about the critical age at which adult mortality is determined from the first articles that were published at the beginning of the 20th century to the most recent research today. With a better understanding of this research, the reader of this monograph will be able to put the results of the following chapters, which mainly deal with the differences in life span by month of birth, into a broader perspective.

Cohort analysis from the 1920s and 1930s for England, Wales and Scotland (Derrick 1927, Kermack, McKendrick & McKinley 1934) suggested that, up until 1925, year of birth was more important than year of death for predicting mortality. In 1927, Derrick plotted age-specific mortality rates for the period 1716 to 1916 against year of birth – rather than year of death – and found that the curves were strikingly parallel. The parallel curves imply that the relative mortality decline of successive birth cohorts is the same in all age groups. Derrick rightly interpreted the parallel curves as indicating that each succeeding cohort experienced a lower mortality risk at all ages. This is contrary to period effects, which would affect different age groups in different birth cohorts and would thus distort the parallelism of the mortality trajectories.

Derrick excluded child mortality under the age of ten because he was interested in the general law of mortality. He argued that the environmental improvements that had resulted in the reduction of child mortality would disturb this general law. Seven years later Kermack, McKendrick and McKinley postulated that it is the level of child mortality that determines the mortality of the cohorts at older ages. In their 1934 article, which has recently been reprinted (Kermack et al. 2001), they used the same data as Derrick but included child and infant mortality. They took the age-specific mortality in the year 1845 for England and Wales as their base year and calculated the relative mortality of the same age groups in the period 1855 to 1925 relative to that base year. They found the following general pattern: that cohorts carried with them the same relative mortality throughout life. At each age, each succeeding cohort had a lower relative mortality than the previous cohort. However, infant mortality did not follow this general pattern and only started to decline after the turn of the century, well after mortality at later ages. The authors argue:

"It has been observed that the improvement in the infantile death-rate became apparent only after 1901. If we remember that before birth and during its first year of life, the child is dependent of its welfare to a very large degree upon the general health and vitality of the mother, then it would be expected that a substantial improvement in the health of the latter would show itself in a reflected improvement in the infantile death-rate. The mothers of 1901 would on the average be born about 1870 or possibly a little later, and, (…), the health of females born at that date has so far improved that the death-rate was improved by about 30 per cent in the case of Scottish and by about 40 per cent in the case of English mothers. It is suggested that this may constitute at least one of the factors conducive to the improvement of infantile mortality in the present century." (p. 681, Kermack, McKendrick & McKinley 1934, 2001)

As an explanation for the regularity of the relative mortality-rates at ages other than infancy and childhood they remark that "the figures behave as if the expectation of life was determined by the conditions which existed during the child's earlier years." And they postulate

"..the health of the child is determined by the environmental conditions existing during the years 0–15, and (…) the health of the man is determined by the physical constitution which the child has built up." (p. 680, Kermack, McKendrick & McKinley 1934, 2001)

The latter two statements obviously contradict the first statement about the onset of the decrease in infant mortality. Since the onset of the decrease in infant mortality is delayed in comparison to the onset of the decrease in adult mortality, it can hardly determine the mortality of the survivors of each cohort.

Despite this contradiction (see also Harris 2001) their paper was a landmark in the analysis of birth cohort influences on adult disease risk since it was the first to formulate hypotheses regarding early-life exposures and their influence on later disease (Davey Smith & Kuh 2001).

Kermack et al. (1934, 2001) performed a similar analysis for Sweden in which they calculated mortality rates relative to the base year 1755. The cohort decline is not as clear in the case of Sweden as it was for England and Wales. The authors argue that a rectangular block dating from 1855 onwards and affecting the age groups 10 to 30 years was the primary cause of disturbances in the cohort decline. If the block is omitted, a cohort decline is observed. This result could not be replicated, however, in a study of Swedish mortality between 1778 and 1993 by Vaupel et al. (1997). The authors calculated the probabilities of death relative to age-specific levels from 1778 to 1799 and plotted them in form of a shaded contour map. They come to the conclusion that "the pattern is clearly more complex than a pure cohort-effect model would suggest" (Vaupel et al. 1997, p. 63). Further studies which took death rates relative to a period earlier than 1870 and which found a pattern that suggests a cohort decline in adult mortality were conducted by Preston and van der Walle (1978) and Coale and Kisker (1990).

In their review of the historical debate on the question of when the mortality risk is determined, Kuh and Davey Smith (1993) show how these early studies influenced public health policies in Great Britain. Public interest in early-life determinants of adult mortality soon dissipated, however. One reason for this was that the year of birth lost its predictive power in terms of mortality prediction by 1925 and period factors became more and more important. Davey Smith and Kuh (2001) updated the table of relative mortality-rates for England and Wales from the Kermack paper and showed that there are dramatic differences between observed and predicted mortality rates. At younger ages, mortality rates fell faster than predicted on the basis of birth cohort regularities, whereas at older ages, mortality declined at a much slower rate than predicted.

Elo and Preston (1992) point out that these early studies were probably so successful in demonstrating cohort effects because they were based on mortality data from a time period before the process of mortality decline on a period level had begun. Once both cohorts and periods began to show mortality improvements it became more difficult to separate the two. Period factors of mortality gained importance as the mortality spectrum shifted from infectious disease towards chronic disease. In the middle of the 19th century, over a third of the deaths in the 15–55 year age group and 15 percent of the deaths at ages 45–64 in Great Britain were due to respiratory tuberculosis (Davey Smith & Kuh 2001). In the United States, TB remained the second leading cause of death at the turn of the century, when it accounted for approximately 11 percent of all deaths (Elo & Preston 1992 citing Mason & Smith 1985). Respiratory tuberculosis is the major cause of death in which childhood conditions are implicated in cohort mortality, and it has received extensive treatment in the analysis of cohort effects on mortality. Davey Smith & Kuh (2001) mention John Brownlee (Brownlee 1916) and the report of the British General Registrar in 1921 (Greenwood 1936) as the first studies that dealt with cohort effects in respiratory tuberculosis. The most frequently referenced article is Frost (1939), where it is pointed out that the changes in mortality from tuberculosis were more consistent on a cohort than on a period basis. Mason and Smith (1985) came to the same conclusion when analyzing the trend in respirator tuberculosis in Massachusetts from 1880 to 1970. Each following cohort had a lower mortality than the previous one, while period coefficients showed little trend with the exception of the advent of successful chemotherapy after 1950 (Elo & Preston 1992). Tuberculosis in late adulthood is believed to originate in infections during childhood. Thus, improvements in earlier life conditions should be reflected in a lower incidence of tuberculosis later in life.

Davey Smith and Kuh (2001) mention stroke, stomach cancer and bronchitis as other important contributors to adult mortality that declined between the mid-19th century and today. For all three of these ailments, there exists evidence of important early-life influences or of a declining birth cohort pattern.

With the shift from infectious to chronic disease in the course of the 20th century, coronary heart disease and lung cancer dominated adult mortality. Coronary heart disease initially shows period-specific rather than cohort-specific increases, and lung cancer shows cohort-specific increases, which reflects the fact that successive groups of young adults take up smoking (Davey Smith & Kuh 2001). Additionally, the introduction of antibiotics for the treatment of infectious disease, which became increasingly effective from the end of the 1930s, influenced mortality in a period-specific rather than a cohort-specific way. Taking all these factors together,

"...the change in the cause-of-death profile (and thus of particular aetiological agents) and the development of medical therapeutics led to change in the relative importance of birth cohort and period effects on all-cause mortality. This change coincided with a shift in focus of public health and epidemiological thinking, from the importance of childhood environment (in particular childhood nutrition) to the importance of adult lifestyles, as major determinants of population health." (p. 696, Davey Smith & Kuh 2001).

It wasn't until the end of the 1970s that the idea of early-life influences on mortality was taken up again. In 1973, Forsdahl published his first paper in a series of five about the effect of very poor living conditions in childhood and adolescence on adult mortality (Forsdahl 1973). This paper, which was originally written in Norwegian, has recently been reprinted in English (Forsdahl 2001). Forsdahl observed that, ever since the start of registration of county mortality in Norway, the mortality of the county of Finnmark has been higher than for the rest of the country. Current socio-economic differences or current differences in life-style behavior like smoking or diet could not explain the difference in mortality. He concluded that the mortality rates, which are 25 percent higher than in other parts of Norway, must be the result of events during childhood and adolescence. To test this hypothesis, he gathered data on infant mortality and adult mortality for the municipality of Sør-Varanger in Finnmark, which is situated between the northern parts of Russia and Norway and is the most remote municipality in the country.

Forsdahl writes in his commentary to the reprint of his paper (Fohrsdahl 2001a) that he was particularly qualified to do the investigation of this region since he grew up in this municipality, where his father was a district doctor and where later, from 1963 to 1974, he became the district doctor. He found that infant mortality in Sør-Varanger was consistent with the infant mortality pattern in Finnmark and that in the past, both had been more than twice as high as in Norway as a whole. Only in the most recent years of his analysis did the infant mortality rates in Norway as a whole and those of Finnmark and Sør-Varanger converge. When he divided the population of Sør-Varanger into people of Finnish and non-Finnish origin he found that both infant mortality and adult mortality are higher for the Finnish than for the non-Finnish population. He observes that, in the past, the Finnish part of the population was generally employed in the more vulnerable branches of the economy such as farming and fishing. At the beginning of the 20th century, Sør-Varanger suffered twice from famine, particularly in regions where the Finnish population is in the majority. In addition, the fertility of the Finnish population was higher and people lived in more crowded environments than the rest of the population. Since the 1950s these differences have disappeared and at the time when he wrote his paper he did not find any difference in current socio-economic and current lifestyle behaviors between the two groups. He concluded his article by observing that

"This study may suggest an association between very poor living conditions in childhood and adolescence and high mortality in adulthood so that the worse the living standards, the higher the later mortality. Thus, one should not expect that the difference in mortality between Finnmark and the rest of the country will disappear until the generation that grew up under these adverse conditions are gone." (p. 307, Forsdahl 2001).

Already in his first paper he mentions that the difference in mortality between the Finnish and the non-Finnish population is mainly due to arteriosclerotic heart disease. Four years later (Forsdahl 1977), he confirmed in an article the hypothesis that poor living conditions in childhood are positively correlated with the risk of dying from arteriosclerotic heart disease. He compares the twenty county rates of all-cause mortality among adults aged 40 to 69 in the period 1964 to 1967 with county infant mortality during the period of their childhood and youth (1896 to 1925). He finds a positive correlation of 0.93 for men and a somewhat weaker correlation of 0.75 for females. This implies that, in counties where infant mortality was high, the same generation had high mortality later in life. The correlation was particularly high and highly significant for arteriosclerotic heart disease and for lung cancer. When he compared adult death rates at county level to current levels of infant mortality, the correlations disappeared. In his next paper he reports a positive correlation between infant mortality and serum cholesterol levels of the same generation at ages 34 to 49 (Forsdahl 1978). In this paper he further develops the explanation given above and suggests "poverty in childhood and adolescence, followed by later prosperity, results in high cholesterol levels."

Forsdahl's analyses have been subject to criticism because he did not control for current socio-economic circumstances, current lifestyle factors or any other current or historical characteristics. He thus cannot exclude the possibility that what he observed was the indirect effects of childhood environment mediated through current life-course conditions. Despite this valid criticism, his studies stand out insofar as they re-introduced the idea that the critical period determining adult mortality is childhood and adolescence. His analyses are the first in a long series of ecological studies that investigated the relationship between infant and childhood mortality and adult mortality of the same generation by geographical area.

The 1980s saw a resurgence of interest in the effects of early-life conditions on late-life mortality which continues up to the present today. Researchers have mainly applied three approaches to demonstrate the importance of early-life conditions. Following the approach of Derrick and Kermack, they tried to separate period and cohort effects; they undertook ecological analysis following Forsdahl's ideas; and they studied the relationship between anthropometric measures, particularly height, and adult mortality. Elo and Preston (1992) reviewed these studies extensively.

1.1 Cohort Studies

Since the 1930s both cohort and period effects have simultaneously influenced old-age mortality. Thus, more sophisticated models were needed to separate the two. Again, Elo and Preston (1992) extensively reviewed those cohort studies that were published until the beginning of the 1990s and this part draws on their results. Caselli and Capocaccia (1989) studied trends in Italian mortality during the 20th century and tried to shed light on the effects of mortality decline in infancy and childhood on a cohort's mortality at older ages. They used age-specific probabilities of dying between the ages of 29 and 79 for the time period 1907 to 1978. These cohorts were born between 1882 and 1953. The cohorts had experienced very different patterns of early mortality – for example, the probability of men surviving to age 15 had increased from 68 percent in 1918 to 82 percent by 1935 (Caselli 1990). The authors estimated logistic regression models of age-specific probabilities of dying which included age, period and a cohort's infant mortality or a cohort's probability of dying during the first 15 years of life as independent variables. Introducing an interaction effect between the latter two variables and age allowed them to study which ages were most affected later in life by changes in early-life mortality. They found that for both sexes higher mortality between the ages of 0 and 15 was associated with higher mortality between the ages of 25 and 45. At older ages, they found the opposite effect, although slightly weaker.

A number of studies (Horiuchi 1983, Wilmoth 1986, Caselli et al. 1987, Vaupel et al. 1987, Wilmoth et al. 1989) took advantage of abrupt changes to identify cohort effects. During the 20th century, the two major wars constituted abrupt changes which affected different cohorts at different ages. A number of these studies found elevated mortality for cohorts who were either born during the two world wars or were in their adolescence at the time. Both male and female children born during World War I in Italy, France, and the former Soviet Union have subsequently experienced higher mortality. Cohort mortality was elevated for German male cohorts whose members were adolescents during World War I. Elevated cohort mortality after World War II has been found for Japanese male cohorts and cohorts of both sexes from the former Soviet Union whose members were adolescents during the war. Both wars affected the subsequent mortality of adolescent cohorts in France and in Italy. Generally, the effect was somewhat stronger for males than females, which suggests that the effects on male cohorts may be due to their participation in the wars at very young ages rather than to nutritional deprivation or poor health conditions.

One example from epidemiological research in which a cohort effect was originally attributed to World War I but is now believed to be rooted in poor living conditions during childhood is peptic ulcer morbidity. Susser and Stein (1962) observed:

"the timing of the first world war, and the unemployment of the 1930s, roughly fit the fluctuations, and the cohorts with the highest peptic-ulcer death-rates were also the chief victims of the first world war. The immediate effects of war are evident in the rise in perforations and deaths from peptic ulcer which followed air-raids and the stress of war. Perhaps in a chronic condition such as this acute events might precipitate prolonged effects". (Susser & Stein 1962, cited by Davey Smith & Kuh 2001)

In 1985 Sonnenberg and colleagues demonstrated that birth cohort dependency of peptic ulcer morbidity could be seen in many countries and that the causative factors became effective before age 15 (Sonnenberg, Mueller & Pace 1985 Sonnenberg 1987). The probable explanation of the cohort effects today is H pylori infection, which occurrs during childhood and is associated with peptic ulcer many years later (Susser 2001). The most recent study analyzes peptic ulcer morbidity and mortality in Iceland (Thors et al. 2002) and cites twelve studies that had demonstrated a cohort pattern previously. The authors find that in Iceland, for all manifestations of peptic ulcer, the disease risk was particularly high in individuals born just after the turn of the century and that it was lower in previous and subsequent cohorts. This time period saw the advent of the fishing industry, which led to overcrowded and poor living conditions. The same cohorts that experienced the highest risk of peptic ulcer disease were also the cohorts with the highest prevalence of H pylori antibodies.

Influenced by the fetal origin hypothesis – which suggests that nourishment before birth and during the individual's infancy programs the development of risk factors for several important diseases of middle and old age – several studies have investigated the effect of famine on cohort mortality. The results, however, are not conclusive. Between 1866 and 1868 a severe famine hit Finland. Kannisto et al. (1997) studied the cohort mortality of more than 600,000 individuals born 5 years before and after the famine and of about 160,000 individuals born during the famine. They found that survival in childhood and adolescence was significantly lower for cohorts born before and during the famine. At subsequent ages, however, including old age, mortality for those born during the famine did not differ from the mortality of those born before or after it.

The effect of the Dutch famine in 1944-1945 on later life disease and mortality is explored in a series of study (Rosenboom et al. 2001, Rosenboom et al. 2000a, Rosenboom et al. 2000b). The authors find that mortality up to age 18 was higher for those born before the famine and those exposed to the famine in the third trimester. Between the ages of 18 and 50, however, no effect of prenatal exposure to the famine could be demonstrated. Results differed for atherogenic lipid profile and heart disease later in life because those who were exposed during early gestation had an increased risk.

Two studies about the long term effects of severe starvation during the siege of Leningrad come to contrary results. (Stanner et al. 1997, Sparén et al. 2003). Stanner et al. did not find a long term effect in 549 children who were born in or near Leningrad during the siege: neither blood pressure, nor glucose intolerance nor lipid concentrations were increased in adult life. In the latter study Sparén et al. find increased blood pressure levels, an increased risk of ischemic heart disease including haemorrhagic stroke among men who experienced the siege at the onset or during puberty They conclude that starvation, or accompanying stress, during adolescence may increase vulnerability to later cardiovascular disease.

Bengtsson et al. (2001) analyzed the effect of environmental stress on old-age mortality in Sweden for the period 1760–1894. They used infant mortality of an individual's cohort to measure the disease load early in life and rye prices to account for nutritional changes. They did not find a significant effect of rye prices during the first years of life on late life mortality, and they also rejected the hypothesis that the disease load on the mother during pregnancy affects the adult mortality of her child. They did, however, find a significant effect of the crude death rate in the first year of life on later-life mortality. The authors concluded that the disease environment in the first year of life rather than nutrition *in-utero* is the critical factor that influences late-life mortality

1.2 Ecological Studies and Place of Birth

Barker and colleagues (Barker & Osmond 1986a, 1986b, 1987, Barker et al. 1989a, Barker et al. 1990, Osmond et al. 1990) performed the most extensive series of ecological analysis to date. Basically, all these studies follow one general research design. They compare adult death rates for specific age groups with infant mortality rates for the same generation on the basis of local administrative regions in England and Wales. In their initial paper (Barker & Osmond 1986a), they found high positive correlations between infant mortality in 1921–1925 and standardized mortality ratios for chronic bronchitis, ischemic heart disease, stomach cancer, and rheumatic heart disease among adults in 1968–1978. For lung cancer and stroke the associations were weaker. They did not correct for any other current or historical conditions. In their next paper – using the same data – they studied the relationship between geographical distributions of infant mortality from bronchitis and pneumonia and current distributions of mortality at ages 35–74 from chronic bronchitis. They found that infant mortality from bronchitis and pneumonia were more closely related with adult mortality from chronic bronchitis in the periods 1959–1967 and 1968–1978 than with any other major cause of death. They controlled for smoking behavior by including lung cancer death rates in their regression model, since smoking is one of the established risk factors for lung cancer. The authors concluded that pulmonary infections in infancy may have persistent effects and raise the risk of death from chronic bronchitis in adulthood. In their next study (Barker & Osmond 1987), they use maternal mortality as an indicator of intrauterine environment and demonstrate a positive correlation between maternal mortality rates in 1911–1915 and mortality from stroke among people aged 55–74 in 1968–1978.

Their next article further develops the idea that the increased susceptibility to cardiovascular disease and chronic bronchitis originates in the intrauterine and early postnatal environment (Barker & Osmond 1989). Barker refers to this study as the first one to show that coronary heart disease results primarily from poor conditions *in-utero* (Barker 2002). Using 212 local authority areas by age at death, the authors found that the geographical distribution of the death rates of stroke at ages 35–74 between 1968 and 1978 are more closely related to neonatal mortality (mortality in the first month) of the same generation than to post-neonatal mortality (mortality between the ages of two months and one year). The reverse is true for ischemic heart disease and bronchitis. This study may thus be seen as the founding study for the fetal origins hypothesis, which states that chronic disease later in life is caused by programming of the infant *in-utero*. It suggests to Barker, that

"...research should be directed towards the intrauterine environment rather than the environment of late childhood-housing, family income, diet, and other influences." (p. 1111, Barker 1990)

The above statement explains why he writes in his commentary to the reprint of Forsdahl's 1993 article:

"Forsdahl's best known paper is entitled: Are poor living conditions in childhood and adolescence an important risk factor for ateriosclerotic disease? The short answer is that they are not." (p. 309, Barker 2002)

However, the ecological studies by Barker and Osmond have been heavily criticized (e.g. Elo & Preston 1992) because the authors failed to control for current socioeconomic conditions, the statistical details presented were insufficient to judge the significance of the reported results, the units and dates of the analyses changed haphazardly from study to study, and statistical controls were inadequate.

In 1991 Ben Shlomo and Davey Smith (1991) addressed the idea that the relationship between past infant mortality rates and present adult mortality simply reflects an unchanging distribution of impoverished socioeconomic circumstances over time. They used infant mortality rates for the period 1895–1903 from 43 administrative county units in England and

Wales and calculated the correlation with current mortality rates with and without correction for current socioeconomic circumstances. They found that the correlations between infant mortality and adult death rates only remained significant for bronchitis, emphysema and asthma when current socioeconomic conditions were corrected for.

Ecological studies of similar research designs have been continued for different countries up to the present (e.g. Elford et al. 1992, Pringle 1998, Maheshwaran et al. 2002, Leon & Davey Smith 2000). Other studies have looked at migrants within a country and whether region of birth contributes to adult mortality independently of the region of residence (Strachan et al. 1995, Schneider et al. 1997a,b, Regidor 2002). Most recently, a survey which was undertaken by Charles Booth at the end of the 19th century in London was digitized using GIS (geographical information systems). The survey portrayed the social and economic conditions and displayed the social classes of inner London on a street-by-street basis. These maps have been digitized, geo-referenced and linked to contemporary ward boundaries. This linkage allowed one to match Booth's measurements of social class with measurements of social class in the 1991 census population and with standardized mortality rates for cause-specific death (Dorling et al. 2000, Orford et al. 2002). An index of relative poverty was calculated for the end of the 19th and 20th centuries respectively. Using partial correlation coefficients, they found that the index of poverty derived from Booth's 19th century observations contributed more to predicting deaths from stroke and stomach cancer in the late 20th century than the index derived from the 1991 census. For other causes of death the modern index contributed more. The authors of the study point out that almost everyone surveyed by Booth's enquiry will have died or moved by 1991. The 19th century index therefore is a truly ecological index and describes area type rather than aggregate characteristics of the resident population. They conclude that the

"ecological associations with past and present deprivation levels of areas do reflect individual level associations of deprivation at different stages of the life course and health outcomes. In short, the longer people spend both in poverty and in poor places, the earlier they tend to die." (p. 1547, Dorling et al. 2000)

The above statement is the antipode of the theory proposed by Barker. To answer the question which of the two hypotheses is true, study designs other than ecological ones were needed. Thus, in the 1990s the emphasis shifted to longitudinal studies on the individual level which included both information about early-life circumstances and life-course information for adult ages.

1.3 The Fetal Origins Hypothesis of Adult Diseases

Since it takes decades for a cohort to age, Barker and his colleagues decided to take a retrospective approach. They traced subjects for whom data are available from their first year of life. They followed up on a historical cohort of 5,654 men born between 1911 and 1930 in Hertfordshire, England (Barker et al. 1989). Information about early-life circumstances was obtained from records kept by attending midwives and local health visitors who made periodic visits during the first year of life. The authors' analysis showed that men with the lowest birth weights and weights at age one had the highest death rates from coronary heart disease and chronic obstructive lung disease. For other causes of death no such trend was seen. Joseph and Kramer (1996) cite 30 studies that followed the same retrospective study design. They investigated the link between measurements of placental weight, birth weight, abdominal and head circumference, ponderal index at birth, weight at age one, infant feeding practices, maternal skinfold thickness and maternal weight gain in pregnancy. The outcomes that were associated with these measurements were clinical phenomena such as blood pressure, serum cholesterol, plasma glucose, plasma fibrinogen, serum concentrations of factor VII and immunoglobulin E, abdominal obesity, and lung function. Direct associations have also been observed between fetal and/or infant growth indices and the prevalence of coronary heart disease, renal complications, and autoimmune thyroiditis. Birth weight, weight at age one and type of infant feeding were correlated with mortality from coronary heart disease, cardiovascular disease, chronic obstructive lung disease, ovarian cancer, and suicide.

Based on these findings it has been hypothesized that physiological or metabolic "programming" occurs at critical periods of early development and substantially determines the occurrence of pathological phenomena later in life. Barker and his colleagues originally proposed this theory in their 1989 paper and have since then further developed it. Most recent studies (Eriksson et al. 1999, Forsen et al. 1999, Eriksson et al. 2001, Forsen et al. 2000, Eriksson et al. 2000) suggest that low birth weight, shortness or thinness at birth, followed by poor infant growth and thereafter rapid growth are associated with coronary heart disease, hypertension and type 2 diabetes. Barker writes:

"People who are small at birth remain biologically different through life because of the persisting constraints and adaptations that accompany slow early growth. Their different morphology and physiology leads them to respond differently to the biological and social environments in later life. Thus the discontinuity in nutritional experience that leads to later disease does not seem to be primarily poor nutrition in childhood and adolescence followed by excess nutrition in adult life (...). Rather it is poor nutrition in-utero and during infancy, the so-called 'fetal' phase of endocrine control of growth, followed by improved nutrition in the second phase of growth, which begins in early childhood " (p. 310, Barker 2002)

One of the strongest and most consistent lines of support for the fetal origins hypothesis of adult disease has been a large series of studies about the relationship between low birth weight, indicating restricted fetal growth, and increased blood pressure later in life. It has been estimated that 1 kg higher birth weight is typically associated with a 2–4 mm Hg lower systolic blood pressure (for a review see Huxley et al. 2002). A recent paper by Leon et al. (2000) demonstrates that among 165,136 Swedish men aged 18 years a 1 kg increase in gestational age-adjusted birth weight was associated with a fall of 1.6 mm HG in systolic blood pressure. However, it has been pointed out (Kramer 2000) that a 1-kg difference in birth weight is enormous (about five or six times the difference in mean birth weight between infants of smoking and non-smoking mothers) and reasonable differences in birth weight are more in the range of a 100g. This makes the effect of birth weight on adult systolic blood pressure levels rather small. In addition, frequent concerns are voiced about confounding by socioeconomic status (e.g. Joseph & Kramer 1996, Kramer 2000).

A recent study by Huxley et al. (2002) casts further doubt on the causal pathway between low birth weight and increased blood pressure. The authors review 55 studies that reported regression coefficients of systolic blood pressure on birth weight. When they weighted each study with the inverse of the variance of the regression coefficient to correct for "statistical size" they found a clear trend: the association between birth weight and blood pressure was weaker in larger studies. It ranged from -1.9 mm HG/kg in those with less than 1,000 participants to -0.6 mm Hg/kg with more than 3,000 participants. Furthermore, in 49 studies which only reported the direction of the association but not the correlation coefficient, only 25 studies found an inverse association. The authors come to the conclusion that the "strong association between birthweight and subsequent blood pressure may chiefly reflect the impact of random error, selective emphasis of particular results, and inappropriate adjustment (...) for confounding factors. These findings suggest that birthweight is of little relevance to blood pressure levels later in life." (p. 659, Huxley et al. 2002)

1.4 Life-Course Factors

The main idea underlying the fetal origins hypothesis of adult diseases is that a critical time period exists early in life. Exposure to deleterious factors during this critical period affects adult health, independently of intervening experience later in life. In an extension of this theory, effect modifiers later in life, e.g. the catch-up growth of low birth weight babies, are recognized. The pathway is predominantly biological.

A life-course approach takes into account that there are mediating factors between the exposure early in life and the manifestation of the disease later in life. A second pathway may be that negative social factors in the early-life environment set individuals onto life trajectories that negatively affect their later health. This pathway is predominantly social. A third pathway is that the cumulative effects of repeated exposure to negative factors adversely affect one's health status either through a biological process (e.g. repeated infections per se) or through a social process (e.g. repeated infections resulting in lower educational attainment).

The life-course approach to chronic disease epidemiology challenges the idea of a critical period and tries instead to disentangle the effect of a critical period from mediating factors of biological or social origin. Ben-Shlomo and Kuh (2002) define the life-course approach as

"the study of long-term effects on chronic disease risk of physical and social exposures during gestation, childhood, adolescence, young adulthood and later adult life. It includes studies of the biological, behavioral and psychosocial pathways that operate across an individual's life course, as well as across generations, to influence the development of chronic diseases". (p. 285, Ben-Shlomo & Kuh 2002)

This definition encopasses the life courses of entire generations and, as the authors mention, is "empirically complex". A large series of studies have been published that fall under this life-course definition (for a review and definition see, e.g., Ben Shlomo & Kuh 2002) many of them dealing with socioeconomic differences early and later in life.

1.5 Month of Birth

How does the indicator month of birth fit into this research? Month of birth is used as an indicator for a critical period early in life. Negative factors during this period affect the metabolism of the body and influence later susceptibility to disease. There is no or little modification of this effect by later-life circumstances and the pathway is purely biological.

Month of birth is not simply another indicator of birth weight. First, birth weight studies usually are based on small samples, obtained from hospital populations. Month of birth is widely available for complete populations, which means that studies based on the month of birth are free of selection biases. Second, birth weight restricts the critical period to the pregnancy of the mother. Month of birth measures seasonal influences during pregnancy and during the first year(s) of life. Third, birth weight is influenced by socioeconomic factors. Thus, the relationship between birth weight and mortality and clinical outcome measures later in life may be confounded by unobserved socioeconomic differences. Birth weight studies definitely require a life-course approach to adjust for possible confounding factors. The month of birth is also not free of socioeconomic confounding. Chapter 3 demonstrates, however, that the social differences, which affect the month of birth, tend to reduce the month-of-birth effect on the life span rather than to strengthen it.

Month of birth has been used extensively in epidemiological research (see Chapter 5), particularly in the area of schizophrenia and other diseases of the mental and nervous system. However, since these epidemiological studies are mainly interested in finding the biological trigger of a particular disease, they have failed to demonstrate the general importance of the month of birth, and thus the early-life environment, on the human life span. If the life expectancy of today's elderly is significantly influenced by their month of birth, then the month of birth must be related to heart disease and cancer, as these are the major causes of death among the elderly. So far, there have been no studies investigating this relationship for heart disease and only very few for cancer.

Differences in life span by month of birth thus demonstrate the importance of the first period of life and the health of the mother during pregnancy, independently of other life-course factors. This monograph shows that the variance in adult life span by month of birth is small compared to the total variance and to the differences among social groups or between men and women. The finding, however, that these differences by month of birth are most probably linked to prenatal or early postnatal conditions related to nutrition or disease is of broad significance and with profound implications for clinical practice and public health policy.

Including this chapter about the epidemiological and demographic context of the month of birth this monograph consists of eight chapters. In the second chapter the differences in life span by month of birth are explored for different populations of the Northern and Southern Hemisphere. The optimal data to test whether a month-of-birth effect exists are longitudinal data where birth cohorts born in a specific season are followed from birth to death. Such data, however, is rare and in most cases only death records are available without information about the population at risk. Therefore, different statistical approaches are used and special attention is given to the question whether the analysis of death data alone introduces a bias into the observed pattern in life span by month of birth. The third chapter is devoted to testing several hypotheses that could explain the month-of-birth effect. A very prominent possible explanation is that the interaction between age and the seasonality in mortality causes the month-of-birth effect. Three different approaches are used to explore this hypothesis. Another explanation is that social differences in the seasonal distribution of births are responsible for the differences in life span. The third hypothesis explores the possibility that children who are born after a certain deadline have to wait another year before they can enroll at school, which makes them approximately a year older than their youngest classmates. This age difference may turn into a lifelong advantage. Finally, selective survival in the first years of life or debilitation in-utero or in infancy may be responsible for the differences observed in life span by month of birth.

The fourth chapter is based on approximately 16 million death records for the United States and explores differences in the month-of-birth pattern in mean age at death by region of birth as well as by education, marital status and ethnicity. The same data is used in chapter five to investigate the month-of-birth pattern for different causes of death. The chapter also contains an extensive review of earlier research that explores the effect of month of birth on diseases and causes of death.

The sixth chapter is devoted to the question whether the month-of-birth effect exists at all ages and whether it is similar for all age groups. Three different data sets – the Danish register data, the Danish twin registry and three consecutive US censuses – are analyzed to answer this question. The choice of the different data sets is motivated by the need to overcome two problems. First, one needs longitudinal data over a long time-period to identify age-specific effects; and second, one needs large data sets to allow for the estimation of age-specific month-of-birth patterns at ages where death rates are low. Since there is no one single data set that combines all these virtues, this chapter resorts to the analysis of several sources that together give a more complete picture.

The seventh chapter explores the month-of-birth pattern of migrants and shows that migrants are subject to the pattern of their region of birth rather than that of their region of residence. It also provides evidence that unobserved socioeconomic characteristics of the migrants may bias their month-of-birth pattern.

The eighth chapter, finally, provides evidence that the month-of-birth pattern in life span is not merely of historical interest but that it may still exist in contemporary cohorts. The monograph closes with a summary and discussion of the evidence presented in the different chapters.

2 Month-of-Birth Patterns in the Northern and Southern Hemisphere

One of the first scholars to be interested in the effect of month of birth on longevity was Ellsworth Huntington. In his book "Seasonality" (1938) he described the relationship between the seasons of the year and social, psychological, and demographic phenomena.

On the basis of 10,890 persons in the Dictionary of American Biography who died at an average age of 68.9 years, he found that those born in February died at age 69.7, whereas those born in June at an average age of 67.8. He observed a secondary maximum in September and October. He obtained similar results for 13,891 New Englanders from genealogical memoirs of 80 families; 3,019 New Englanders and New Yorkers; 12,173 people from New York, New Jersey and Pennsylvania and 9,921 people from Maryland, Virginia, North and South Carolina, a few from states farther south or west; and 7,517 people in the British "Who Was Who".

Huntington comes to the conclusion that, in all populations (with the exception of the British population), people born in February or March live longer than those born in July or August. Huntington's finding stimulated extensive research in the area of psychology. Since the 1930s more than 200 studies have been conducted about the prevalence of schizophrenia and month of birth (for a review see Torrey 1997). More recently, researchers attempted to attribute the differences in the prevalence of schizophrenia by month of birth to maternal virus infections such as measles or flu during pregnancy. The results, however, are not conclusive.

Eysenck took up the topic of the impact of month of birth on life span in his book "Astrology. Science or Superstition?" in 1982 (Eysenck & Nias 1982) and Miura edited a series of epidemiological studies in his book entitled "*Seasonality*" (Miura 1987). In the 1980s and 1990s quite a few epidemiological studies were conducted that found a relationship between month of birth and specific diseases such as allergies, insulin-dependent diabetes, congenital malformations, Parkinson's and Alzheimer's disease, breast cancer, etc. (see Chapter 5). In all these studies month of birth is used as an indicator for the environment early in life.

This chapter takes up the topic and provides evidence from population data that month of birth and life span are indeed related. In contrast to earlier research, this study is not based on sample data but on complete population data for countries of the Northern and Southern Hemisphere.

Huntington conjectured that the month of birth is an indicator for environmental factors that are linked to the seasons of the year. If this is true, then the patterns of two geographically close populations should resemble each other, and the pattern in the Northern Hemisphere should be mirrored in the Southern Hemisphere. Furthermore, life spans of people who were born in the Northern Hemisphere but who died in the Southern Hemisphere should resemble the pattern of the Northern Hemisphere.

The first step, therefore, was to obtain data on the populations of Denmark, Austria, and Australia to test this conjecture. Statistics Denmark provided longitudinal data based on the Danish population register that follows every person living in Denmark from 1968 to the present. Statistics Austria and Statistics Australia provided death certificates for all deaths that occurred between 1988 and 1996 in Austria, and between 1993 and 1997 in Australia. In addition to these three countries, data were also obtained for Hawaii. Hawaii is of particular interest for the study of differences in life span by month of birth because it is located close to the equator. The data for Hawaii are based on US death records for the years 1989 to 1997; these records include place of birth.

The optimal data to test for differences in life span by season of birth are longitudinal data. Birth cohorts born in a specific season are followed from birth to death and life expectancy can be calculated using simple life-table methods. Such data rarely exist however. The data that are closest to this requirement are register data from the Scandinavian countries. The Danish data used in this study consist of a mortality follow-up of all Danes who were at least 50 years old on 1 April 1968. This is a total of 1,371,003 people, who were followed up to week 32 of 1998. The study excludes 1,994 people who were lost to the registry during the observation period. Among those who are included in the study, 86% (1,176,383 individuals) died before week 32 of 1998; 14% (192,626 individuals) were still alive at the end of the follow-up.

Thus, for Denmark both the risk population and the number of deaths are known, which means that it is possible to estimate remaining life expectancy at age 50 on the basis of life tables that were corrected for left truncation. This was achieved by calculating occurrence and exposure matrices that take into account an individual's age on 1 April 1968. For example, a person who was 70 at the beginning of the study and who died at age 80 enters the exposures for ages 70 to 80 but is not included in the exposures for ages 50 to 69. The central age-specific death rate is based on the occurrence-exposure matrix. The corresponding life-table death rate is derived by means of the Greville Method (Greville 1943).

Population registers do not exist for Austria, Hawaii, and Australia, where only individual death records are available. Exact dates of birth and death are known for a total of 681,677 Austrians who died between 1988 and 1996 and for 219,820 native-born Australians who died between 1993 and 1997 at ages 50+. 42,969 decedents of similar age were born in Hawaii and died between 1989 and 1997. The population at risk, however, is unknown, which means that life span by month of birth cannot be estimated on the basis of simple life-table techniques. For Austria, Australia, and Hawaii remaining life span at age 50 was therefore estimated by calculating the average of the exact ages at death.

Mean age at death is not equivalent to life expectancy when cohorts are not extinct. Gavrilov and Gavrilova (2003) pointed out that mean age at death is influenced by changes in the seasonal distribution of births. In Northern Europe the number of births generally peaks in February and March and reaches a minimum in December. Suppose that in younger cohorts the seasonality in births has become smaller as compared to older cohorts. This would imply that in younger populations there are proportianally less people born in winter and more born in fall. In death data this shift in the birth distribution will be reflected in a higher mean age at death for those born in winter and a lower mean age at death for those born in fall.

The usual procedure to account for the effect of possible shifts in the seasonal distribution of births is to compare the seasonal birth distribution at the time of birth of a cohort with the birth dates of the deceased or the survivors of the birth cohort at a given age. Unfortunately, for Australia and Hawaii this information is not available. For Austria the seasonal distribution of births is recorded in statistical yearbooks for the years 1881-1912 and it will be compared with the birth dates of the survivors in the 1981 census. The resulting pattern is compared with the month-of-birth pattern on the basis of the mean age at death.

Furthermore, the Danish twin registry is drawn on for the calculation of remaining life expectancy using the information about the population at risk and the number of deaths. The month-of-birth pattern is then compared with the pattern on the basis of the death counts alone.

In all these analyses the age range is generally restricted to ages 50 and above so as to exclude the possibility that the differences in life span simply reflect differences in survival during the first part of life by month of birth rather than at old age.

2.1 Life Span by Season of Birth

A similar relationship between month of birth and life span exists in both of the Northern Hemisphere countries. Adults born in the autumn (October–December) live longer than those born in the spring (April–June). The difference in life span between the spring- and autumn-born is twice as large in Austria (0.6 years) as in Denmark (0.3 years).

In Denmark remaining life expectancy at age 50 is 27.52 years. The average age at death is lowest among those born around week 18, and it peaks at week 51. The life span of Danes born in specific months varies periodically around the mean (Fig. 2.1.C). For those born in the second quarter, life spans are 0.19 ± 0.05 years shorter than average; for those born in the fourth quarter they are 0.12 ± 0.04 years longer than average. This difference is statistically significant (Cox-Mantel statistic: $p<0.001$).

In Austria deaths occurred at an average age of 77.7. The mean life span of people born in specific months of the year deviates from this average (Fig. 2.1.A). The average age at death is lowest for those born around week 20 and highest for those born around week 46. The deviation in mean age at death is highly significant (Bonferroni test: $p<0.001$) for those born in the second and the fourth quarters. The life spans of people born between weeks 14 and 26 are 0.28 ± 0.03 years below average; life spans of those born between weeks 40 and 52 are 0.32 ± 0.03 years above average.

The pattern in the Northern Hemisphere is mirrored in the Southern Hemisphere. The mean age at death of people born in Australia in the second quarter of the year is 78.0; those born in the fourth quarter die at a mean age of 77.65. The difference of 0.35 years is statistically significant (Bonferroni test: $p<0.001$) (Fig. 2.1.B).

Significant differences in the life span by month of birth exist for Hawaii (Fig. 2.1.D). The mean age at death of people born in Hawaii is 74.5. The mean age at death of the March-born is 0.53 years above the average; that of the October-born is 0.29 years below the average. The difference between the two months is statistically significant at $p=0.004$ (Bonferroni test).

Based on the results reported above one can conclude that the pattern in the Southern Hemisphere is a reversed mirror image of the Northern Hemisphere pattern. This can be clearly seen from the correlations of the mont-of-birth patterns. The correlation of the deviations in life span by month of birth between Austria and Denmark is 0.83 (Pearson correlation, one sided test: $p<0.0001$). Between Austria and Australia it is -0.79 ($p<0.001$) and between Denmark and Australia -0.80 ($p<0.001$).

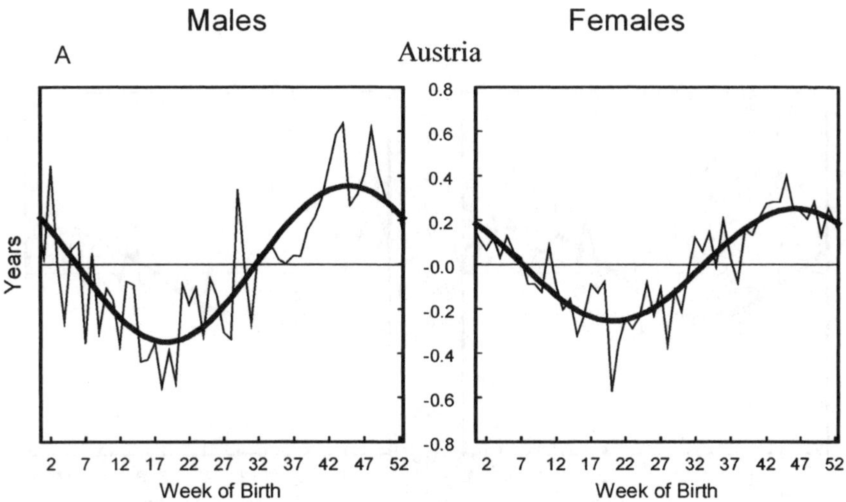

Figure 2.1. Austria (A) and Australia (B): Mean ages at death by week of birth. The fitted line is estimated by a cosine term with a period of 52 weeks.

Single cosinor analysis with a period of 52 weeks (or 12 months) shows that in all four populations the differences in lifespan follow a seasonal pattern. For each of the four populations, ages at death (Austria, Australia and Hawaii) and remaining life expectancy at age 50 (Denmark) are fitted

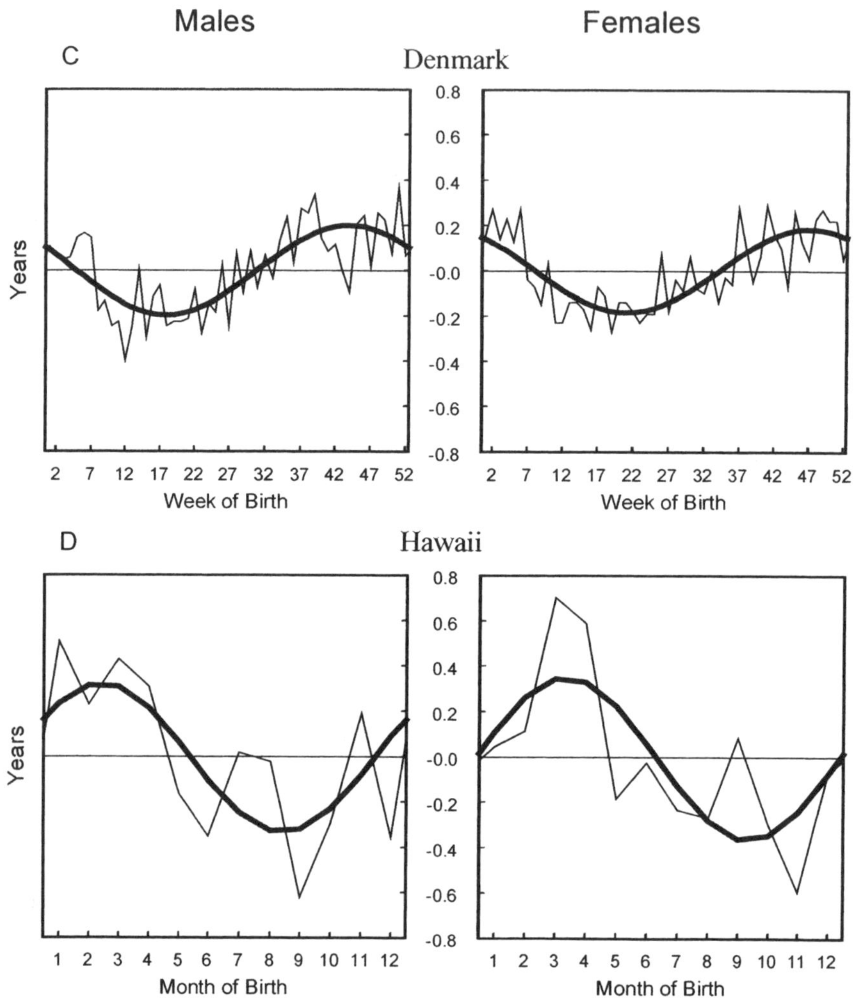

Figure 2.1. (continued) Denmark (C) and Hawaii (D): Mean ages at death by month of birth. The fitted line is estimated by a cosine term with a period of 12 months (Hawaii) and 52 weeks (Denmark).

by a cosine term $age= a_0*cos(t-a_1)$, where a_0 is the amplitude, a_1 the acrophase (maximum), and $t=week\ of\ birth\ /52*2\pi$ (Hawaii: $t=month\ of\ birth\ /12*2\pi$). The parameters are estimated using the least squares estimation procedure. Figures 2.1.A–D show the weekly means together with the fit-

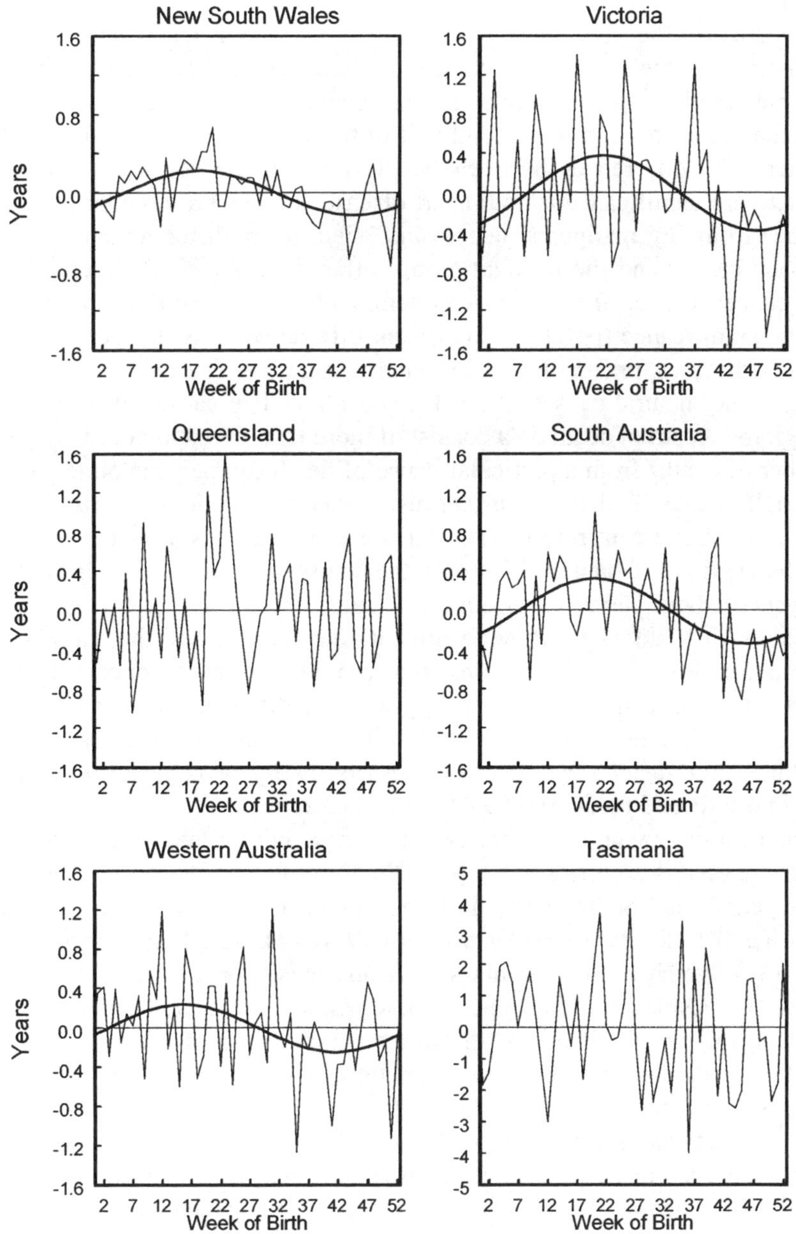

Figure 2.2. Australia: Mean age at death by week of birth and region of birth. The fitted line is estimated by a cosine term with a period of 52 weeks. The cosine term is significant for all regions with the exception of Queensland and Tasmania.

ted cosine terms. The fit of the sinusoidal functions is highly significant, with $p<0.001$.

For Austria the data allow us to analyse causes of death for the years 1990-1997 (Table 2.1). There is a general tendency for people born in the first half of the year, especially in the second quarter, to die at younger ages than those born in the second half of the year, especially in the fourth quarter. The differences between people born in the second and fourth quarters are significant for both heart disease and cerebrovascular disease. In the group of malignant neoplasms, significant differences exist for stomach cancer and the residual group 'other neoplasms'. Differences for lung cancer, cancer of the urinary system, and diabetes mellitus are of borderline significance ($p=0.06$). Significant differences exist for chronic respiratory diseases, pneumonia and influenza, digestive diseases, the residual group 'other natural causes of death', and for violent causes of death. Although the Austrian death data consist of more than 600,000 decedents, the number of deaths from a particular cause of death can become quite small, especially when the data are further distinguished by month of birth. Some of the non-significant results may thus be caused by insufficient numbers of observations. Chapter 5 reports the cause-specific results from the analysis of about 16 million death certificates from the United States.

Since Australia is a whole continent, the analysis of the month-of-pattern for different birth regions may provide evidence concerning the underlying causal mechanisms. It appears that the reversal of the pattern seen in the Northern Hemisphere exists in almost all Australian states and territories: the mean age at death in the first half-year is generally higher than in the second half-year (Table 2.2, Fig. 2.2).

The peak-to-trough difference between those born in the second and the fourth quarter is significant for New South Wales (0.39 years), Victoria (0.89 years), and South Australia (0.61 years). It is of borderline significance ($p<0.1$) for Western Australia (0.49 years). The largest difference ($p=0.157$), which is however not significant, exists for Tasmania, with 1.6 years. The cosine analysis supports these results. The cosine functions are generally highly significant with the exception of Tasmania and Queensland. The regional differences in the pattern of life span by month of birth are not significant (ANOVA F-test for the interaction effect between month of birth and region: $p=0.585$).

The most interesting result of the regional analysis is that the peak-to-trough difference for Queensland is not significant. The non-significant result is not due to small numbers of observations. There are more observations for Queensland than, for example, Victoria, where the difference is highly significant. Figure 2.3 reveals that in Queensland, like in the other

Table 2.1. Difference in mean age at death for people born in a specific season from average age at death by major causes of death; Austria 1990-1997.

Causes of death (ICD Number)	Season of birth					Number of deaths
	Winter (Jan-Mar)	Spring (Apr-Jun)	Summer (Jul-Sep)	Autumn (Oct-Dec)	p*	
Heart disease (390-429, 440-458)	-0.08	**-0.30**	0.06	**0.33**	0.00	257,167
CVD (430-438)	-0.01	**-0.23**	-0.03	**0.27**	0.00	80,367
Malignant neoplasms						
Breast (174)	0.12	**-0.25**	0.00	**0.14**	0.53	12,201
Uterus & female genital organs (179-184)	0.05	**-0.08**	**0.10**	-0.07	0.92	9,530
Prostate & male genital organs (185-187)	**-0.18**	-0.05	0.01	**0.24**	0.40	9,401
Urinary system (188-189)	-0.09	0.07	**-0.39**	**0.42**	0.06	8,508
Haemoblastoses (201-208)	**-0.19**	-0.07	-0.02	**0.28**	0.44	9,147
Lungs (162)	-0.14	**-0.18**	0.06	**0.26**	0.06	24,178
Stomach (151)	-0.10	**-0.44**	0.03	**0.50**	0.00	12,689
Intestines (152-154)	-0.02	**-0.12**	**0.12**	0.02	0.70	21,233
Other	-0.07	**-0.24**	-0.01	**0.33**	0.00	39,359
Respiratory system (460-479, 488-519)	0.04	**-0.18**	-0.16	**0.33**	0.05	18,018
Pneumonia and influenza (480-487)	-0.01	-0.10	**-0.35**	**0.45**	0.02	10,807
Digestive system (520-579)	0.23	-0.20	**-0.28**	**0.24**	0.01	27,384
Diabetes mellitus (250)	-0.06	**-0.33**	0.19	**0.20**	0.06	13,537
Other natural	**-0.26**	-0.22	-0.08	**0.58**	0.00	27,197
Violent (E800-E999)	-0.04	**-0.41**	-0.01	**0.49**	0.00	21,998

Bold figures: maximum and minimum difference from average age at death

* p value: Anova F-Test for all seasons

Table 2.2. Mean age at death by state of birth and quarter of birth, amplitude, and peak estimated by cosinor analysis based on Australian death records for the years 1993 to 1997.

	Quarter of Birth				Max. Dif.+	Cosinor Analysis		Deaths
	1	2	3	4		Amplitude	Maximum Week	
New South Wales	77.1	**77.2**	**77.0**	76.9	0.39**	0.23 0.14-0.31	19 16-22	126,308
Victoria	**77.1**	**78.2**	78.1	77.3	0.89**	0.38 0.13-0.96	22 16-27	14,130
Queensland	**77.4**	**77.7**	77.5	77.6	0.31	n.s.	n.s.	20,686
South Australia	78.3	**78.5**	78.2	**77.9**	0.61**	0.33 0.17-0.49	20 16-24	32,260
Western Australia	**77.5**	77.4	77.2	**77.0**	0.49**	0.24 0.05-0.43	16 9-22	23,395
Tasmania	78.6	**79.5**	78.3	**77.9**	1.60	n.s.	n.s.	1,843

** Bonferroni test: $p<0.01$, * Bonferroni test: $p<0.1$, bold figures indicate the minimum and the maximum mean age at death

\+ Maximum difference in mean age at death between quarters

n.s. cosinor term not significant

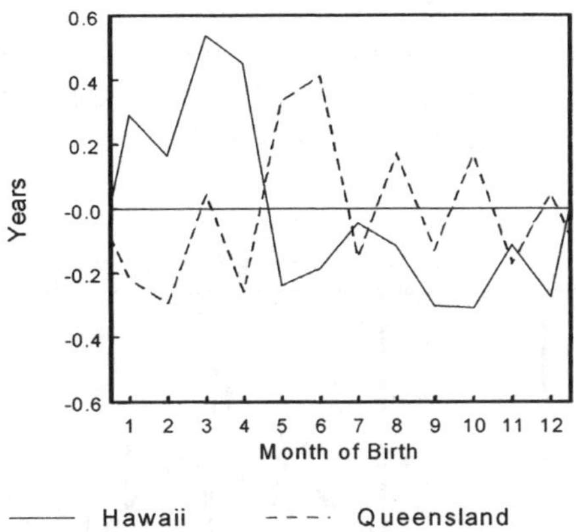

Figure 2.3. Queensland and Hawaii: Deviation in life span for people born in a specific month from the average remaining life span at age 50.

Australian states, mean age at death is highest for those born in May or June. It is lowest, however, for those born in January and February. The two major cities of Queensland are Brisbane (27.39 South Latitude, 153.12 East Longitude) and Townsville (19.25 South Latitude, 146.77 East Longitude). Queensland is geographically almost a mirror image of Hawaii, whose capital, Honolulu, is situated at 21:18 North Latitude and 157:51 West Longitude. Comparing the month-of-birth patterns for these two countries one finds that they are reversed: in Hawaii mean age at death is highest for those born during the first four months, in Queensland it is lowest; from May on mean age at death is generally lower in Hawaii and generally higher in Queensland. The correlation between Hawaii and Queensland is -0.594 and significant at p=0.021.

Comparing the seasons of the year in both Hawaii and Queensland, those born at the beginning of the winter season, which is characterized by lower temperatures, live longer. In tropical northern Australia the wet season corresponds with summer and lasts from November through April. The dry season corresponds with winter and lasts from May through October. In Hawaii the wet season coincides with moderate temperatures and lasts from October to May, while the hot and dry summer lasts from June to September.

The Australian data distinguish between native-born Australians of foreign heritage and aborigines and contains the death records of 2,254 Aborigines and 218,279 native-born non-indigenous Australians. These figures exclude records with birth dates 1 January and 1 July because of a heaping of birth dates on these two days. It is highly likely that unknown birth dates were assigned randomly to one of these two dates. The mean age at death of aborigines is almost 10 years lower (67.9 years) than that of the non-indigenous Australian population (77.8 years). The week-of-birth pattern is similar (Fig. 2.4), but the amplitude is much larger among aborigines (1.17 years, 95% CI: 0.40-1.94). The cosine function peaks in week 20 (95 % CI: 15-25). The amplitude among the native-born non-indigenous Australian population is 0.24 years (95% CI: 0.18-0.30), mean age at death peaks for decedents born in week 19 (95% CI:17-22). The fit of the cosine functions is highly significant at p<=0.001.

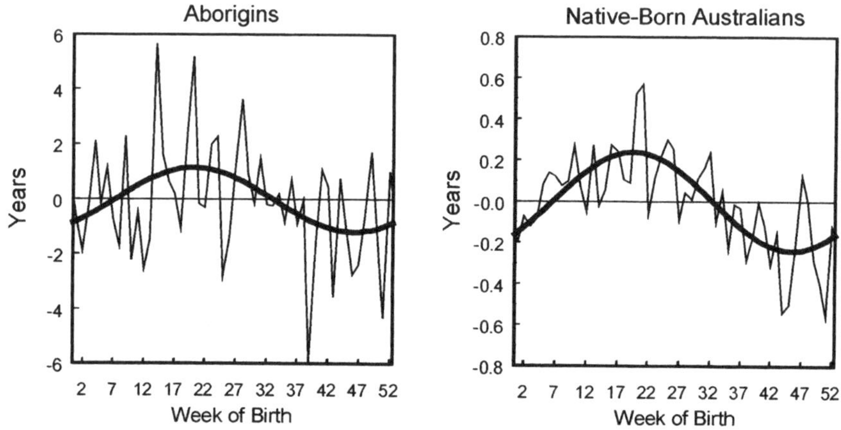

Figure 2.4. Aborigines and native-born Australians of foreign heritage: Deviation in mean age at death for people born in a specific week from the average age at death (ages 50+) estimated by a cosine function with a period of 52 weeks.

2.2 A Different Approach

If people born in a specific month experience a higher mortality risk than others, the distribution of birth dates of the total population changes with age (Vaupel & Yashin 1985). This can be easily tested on the basis of population censuses. A simple tabulation of the Austrian 1981 census gives the proportion of those born in the first and fourth quarters for five-year age-groups. A lower mean age at death for people born in the second and third quarters implies that, with advancing age, the proportion of people born in the first and fourth quarters of the year will increase. This is exactly what is to be observed in the Austrian 1981 census, where the proportion increases by almost 5 percentage points from 49.7 per cent at ages 50-54 to 54.5 per cent at ages 95+. The change in the proportion is statistically significant at p=0.0001 (χ^2 test).

This change could also be due, however, to a change in the seasonal distribution of births. For example, it has been shown that, in Austria, the seasonal fluctuations in the number of births changes between the periods 1881-1912 and 1947-1959 (Doblhammer et al. 2000). A better approach, therefore, is to compare the seasonal distribution of births of a cohort with the seasonal distribution of birth dates among survivors fifty or more years later. For Austria the seasonal distribution of births is available for almost all years between 1881 and 1912. The statistical yearbooks, which were published annually by the Central Bureau of Statistics of the Austro-Hungarian Empire, contain the number of births by month of the year. It is therefore possible to compare the seasonal birth distribution of the years 1881 to 1911 in the German-speaking regions of the Austro-Hungarian Empire with the distribution of the survivors in the 1981 Austrian census. At the time of the census the survivors were between the ages of 70 and 100 (Table 2.3). The distribution of birth dates is clearly different at the time of birth and in the 1981 census. For all birth cohorts the proportion of the fall-born increases with age. The increase is 1.1 percentage point for the youngest cohort (1907-1911) and 4.4 percentage points for the oldest cohort (1881-1886).

Table 2.3. Percent of winter births at the time of birth and in the Austrian 1981 census for different birth cohorts.

Birth cohort	Age of the cohort in the 1981 census	% winter births at the time of birth	% winter birth dates in the 1981 census	% Difference
1881-1886	95-100	50.1	54.5	4.4
1887-1891	90-94	50.2	53.8	3.6
1892-1896	85-89	49.9	53.3	3.3
1897-1901	80-84	49.8	52.4	2.6
1902-1906	75-79	49.9	51.4	1.5
1907-1911	70-74	49.7	50.8	1.1

2.3 Bias in the Month-of-Birth Pattern

In the case of extinct cohorts, mean age at death is similar to life expectancy estimated by the life-table method. The data that exist for Austria,

Australia, and Hawaii, however, do not come from extinct cohorts and mean age at death is therefore a biased estimate of life expectancy.

One unobserved factor that causes bias in the month-of-birth pattern is a change in the seasonal distribution of births over time. In Northern Europe the number of births usually peaks in February and March, declines thereafter, reaches a secondary peak in September and a trough in December. In the United States the pattern is reversed with a trough in spring and a peak in late summer and fall (Lam & Miron 1996). The basic seasonal pattern has not changed over time but in some countries such as the United States the fluctuations became smaller between 1947 and 1976 (Seiver 1985), in other countries such as Austria they became larger (Doblhammer et al. 2000).

Take the case of Austria, where information about the monthly number of births exists for the years 1881 to 1912 and 1945 onwards. No information is available starting from World War I, which led to the collapse of the Austrian Hungarian Monarchy and the destruction of most administrative structures, until the end of World War II.

The seaonal pattern develops as follows: the peak in February decreases from 8.83% for birth cohorts 1881-1890 to 8.73% for birth cohorts 1891-1900. It remains stable for birth cohort 1901-1912 and considerably increases (9.15%) for birth cohorts 1947 to 1959. The intermediary September peak sharply increases between 1881-1990 (8.08%) and 1891-1900 (8.33%) and remains stable thereafter, while the December trough remains almost unchanged between 1881-1890 and 1947-1959. These changes in the seasonal distribution of births result in proportionally more younger decedents among the February-born, and mean age at death will therefore underestimate true life expectancy.

In the following an attempt is presented for assessing the bias by using the information about the seasonal birth distribution for the years available. In the Austrian death records of the years 1988 to 1996 decedents aged 50 and above were born between 1881 and 1946. The frequency distribution of birth years is: 1881-1890: 0.14%, 1891-1900: 7.2%, 1901-1912: 44.3%, 1913-1922: 25.4%, 1923-1932: 16% and 1933-1946: 7.1%. To calculate the number of expected decedents by month of birth, one needs the average seasonal birth distribution for the time period 1881-1946. This average is derived by weighting the seasonal birth distributions for the respective time period with the according frequencies of birth years observed in the death records. Since no seasonal birth distribution is available for the time periods 1913-1922, 1923-1932 and 1933-1946, it is assumed that the seasonal distribution of the period 1901-1912 holds true for the period 1913-1922, while from 1923 onwards the seasonal distribution of births follows the pattern observed in 1947-1959. The distribution of birth months in the death records is then compared with the distribution of birth months in the average seasonal birth distribution. A similar calculation is performed for all decedents aged 80+. In the death records they were born between 1881 and 1916 and the seasonal distribution of the number of births is known with the exception of the last 4 years. Again a weighted seasonal birth distribution is calculated and compared with the distribution of birth months in the death records.

Figure 2.5 shows the month-of-birth pattern based on mean age at death, together with the percentage deviation of the birth months in the death records for ages 50+ and 80+ as compared to the average seasonal birth distribution for the time period 1881-1946 (ages 50+) and 1881-1916 (ages 80+). For example, the value -0.06 indicates that for ages 80+ the proportion of the February-born is 0.06 percentage points lower in the death records than in the corresponding seasonal birth distribution of the years 1881-1916.

Comparing the two trajectories in Figure 2.5.A it appears that the pattern based on mean age at death is shifted to the left for those born in the first half-year. In other words, excess mortality, particularly of those born in the first three months, is overestimated. This is especially true for ages 50+. For ages 80+ the patterns based on mean age at death and on the frequency distribution of birth dates are close, and no serious bias is introduced by using mean age at death as a measure of life expectancy.

The seasonal birth distribution for the period 1881-1912 is based on the

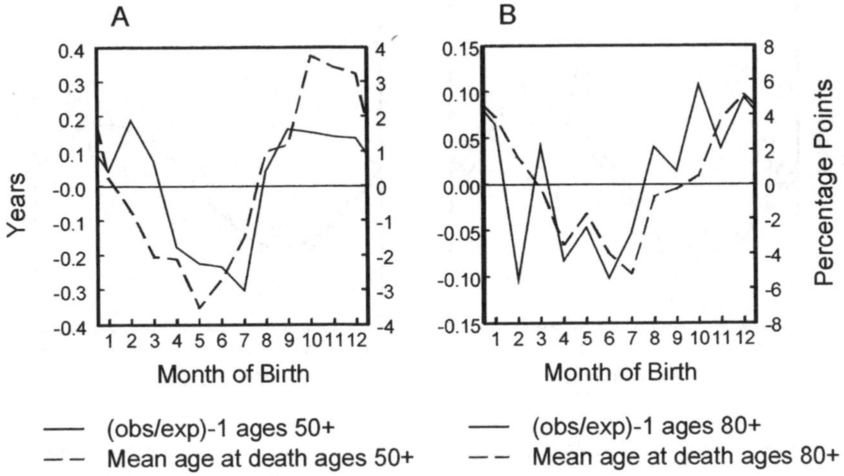

Figure 2.5. Deviation of mean ages at death from average mean age at death for ages 50+ (A) and 80+ (B) and percent deviation of the observed seasonal birth distribution for ages 50+ and 80+ from the expected distributions. Austrian death records 1988-1996 and seasonal distribution of births for the years 1881-1912 and 1947-1959.

German speaking core of the Austrian Hungarian Empire, which is largely identical with the area which constituted the Austrian Republic from 1918 onwards. A considerable proportion of the decedents that were born before 1918, however, was born outside of this core-region.

The Austrian death records do not include the place of birth, they only distingiush between Austrians and foreign nationality. Decedents born before 1918 in regions of the Monarchy that were outside the German speaking core are considered Austrians. An unpublished review of the seasonal birth distributions in the different regions of the Monarchy shows that the distributions differ enormously. It is therefore impossible to determine the exact seasonal birth distribution that should be used for the comparison with the death records.

The second data source that can be used for assessing the difference in the month-of-birth pattern in life span on the basis of mean age at death and on life expectancy is the Danish twin registry. The Danish twin registry was established in 1954 and contains all twin pairs born between 1870 and 1910 and all same-sex pairs born between 1911 and 1930.

Twin pairs were included when both partners survived their 6th birthday. Birthdates of 28,102 twins are included in the registry. Of these, 4323 twins are of unknown survival status, 7,368 twins are censored due to emi-

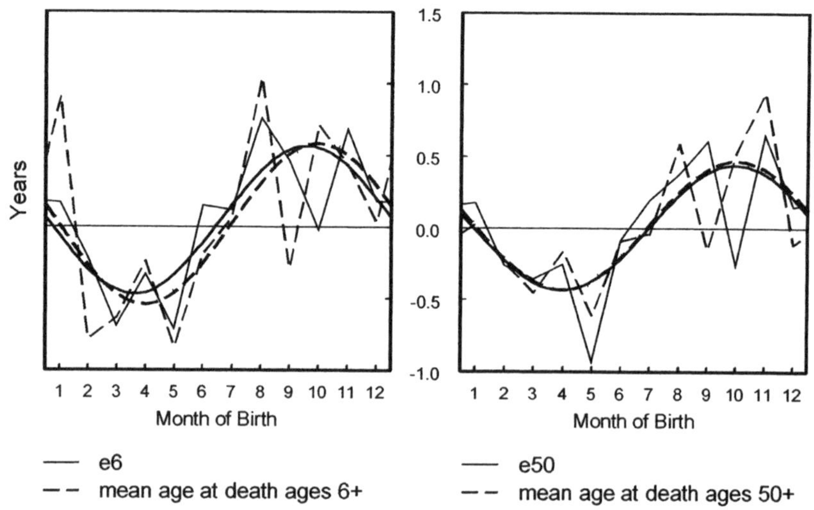

Figure 2.6. Deviation in mean age at death (ages 6+ and 50+) from average mean age at death and deviations in life expectancy (e^6 and e^{50}) from average life expectancy by month of birth. The bold sinusoidal functions are fitted cosine terms.

gration or survival at the end of the observation period (September 1995), 16,411 twins have died.

The frequency distribution of birth cohorts is shifted towards the more recent cohort and is as follows: 1870-1879: 8.5%; 1880-1889: 11.7%; 1890-1899: 13.8%; 1900-1909: 17.9%; 1910-1919: 22.0%; 1920-1930: 26.1%. As expected, life expectancy and mean age at death differ widely: remaining life expectancy at age 6 is 76.87 years, mean age at death 67.24 years. The difference remains for ages 50+ with a remaining life expectancy of 78.52 years and a mean age at death of 73.74 years. The month of birth pattern, however, is almost unchanged (Figure 2.6) and neither the peak to trough difference nor the pattern itself depends on whether life expectancy or mean age at death is used in the calculation.

The two examples show that using mean age at death from death records of not extinct cohorts results into a minor bias concerning the month-of-birth pattern in the case of Austria and into no bias at all in the case of Denmark. For populations without information about the seasonal distribution of births and without population registries mean age at death therefore gives a reasonable approximation of the month-of-birth pattern.

3 Five Hypotheses

The previous chapter explores whether differences in life span after age 50 by month of birth exist in the four populations Austria, Australia, Denmark and Hawaii. This chapter is devoted to testing five hypotheses. The first hypothesis assumes that the interaction between age and the seasons of mortality causes the differences in life span by month of birth. For example, people born in April are older than people born in November when the high mortality of winter strikes them. The second hypothesis tests whether the differences are due to unobserved social factors that influence or result from the seasonal timing of births. The third hypothesis considers whether deadlines for school attendance may have an impact. The fourth hypothesis explains the differences in adult life span by differential survival in the first year of life, while the fifth hypothesis assumes that debilitation *in-utero* or in the first year of life increases the infant's susceptibility to diseases at adult ages.

3.1 Seasonal Distribution of Deaths

The first hypothesis assumes that the observed relationship is caused by the interaction of the seasonal distribution of deaths and the monthly increase in adult mortality. For example, since the April-born are older than the autumn-born when the high mortality risk of winter strikes them they may experience higher mortality overall. In other words, month of birth is not so much related to the beginning or life but is rather a life-cycle indicator. This hypothesis has already been widely discussed and tested in the research about schizophrenia. In this research it is referred to as the "age prevalence and age incidence effect". Torrey et al. (1997) review a large number of studies that deal with this "statistical artifact", and its description here is largely based on Torrey's article. The studies about age prevalence and age incidence deal with the problem of the interaction between age and seasonality in the incidence of schizophrenia, although the term seasonality is not mentioned explicitly. Lewis and Griffin (1981, p. 590) explain the effect as follows:

"For example, people who are 30 years old have a higher incidence of schizophrenia than people who are 29 years old. Therefore, people born in 1937, may have a higher incidence in 1967 than in 1966. When first admissions from 1967 are pooled with first admissions from 1966, people born in 1937 and admitted in 1967 will have more influence on the final result than people born in 1937 and admitted in 1966. The difference in the influence between any 2 years is not great, but over the course of an entire data set it becomes noticeable and should be corrected."

Lewis and Griffin try to explain – in a somewhat complicated manner – that the incidence of schizophrenia is seasonal and that people born in different seasons of the year are hit by the higher risk of developing schizophrenia at slightly different ages. This is similar to the problem one faces with mortality.

Torrey et al. write that although "the age incidence and the age prevalence effects have elicited extensive and spirited discussion among statisticians and psychiatric researchers (...), the effects have been shown to be negligible." Hare (1975) and Lewis and Griffin (1981) had proposed a formula to correct for the effect. Torrey et al. mention thirteen studies that used the correction but still found significant winter–spring birth excess among schizophrenic patients.

In Denmark, Austria, and Australia the risk of death is lowest in the summer and highest in the winter. The differences between the summer trough and the winter peak are considerable, and they depend on the time period, age, climate, and social conditions. In all three countries death rates after age fifty increase by about 0.8% per month of age. The question is whether the increasing mortality with age together with the seasonal differences in mortality may cause the differences in life span by month of birth.

In the following three approaches are used to explore the impact of the seasonal distribution of deaths on the month-of-birth pattern in life span. The first approach is merely descriptive and displays the death rates by month of birth, current month and age in month. The second approach estimates the age-specific force of mortality depending on the month of birth and the current half-year. The third approach models the age-specific death rates by month of birth, current month and age in months.

3.1.1 A Descriptive Approach Based on Death Rates

In the following, a descriptive approach is used for a first analysis of the season of birth effect. On the basis of the Danish longitudinal data, the population at risk and the number of deaths by month of birth and age in months are calculated. Let x denote age in integer years and let y denote the current year. Let i denote age in months since last birthday, j the current month, and k the month of birth; let D_{ijk}^{xy} be the number of deaths that occur for people of age $x+i/12$ years at time $y+j/12$, and let T_{ijk}^{xy} be the corresponding size of the exposed population at risk of death. Note that the month of birth k is already defined by the combination of current month j and age in months since last birthday i. Let n_j denote the number of days in current month j. When calculating the exposed population, all deaths were assumed to occur in the middle of each month. The relative deviation of the monthly death rate from the annual death rate, R_{ijk}^{xy}, was calculated as

$$R_{ijk}^{xy} = \frac{D_{ijk}^{xy}}{T_{ijk}^{xy} n_j} \Big/ M_{xy} \qquad [3.1]$$

where

$$M_{xy} = \frac{\sum_{i=1}^{12} \sum_{j=1}^{12} D_{ij}^{xy}}{\frac{365.25}{12} \sum_{i=1}^{12} \sum_{j=1}^{12} T_{ij}^{xy}}. \qquad [3.2]$$

Let M denote the average death rate D/T over the whole observation period and over all ages. Summing the number of deaths and the population at risk over age x, time y, current month j and age since last birthday i and dividing it by the average death rate M then gives R_k (Equ. 3.3) the death rates by month of birth as compared to average mortality.

$$R_k = \frac{\sum_x \sum_y \sum_i \sum_j D_{ijk}^{xy}}{\sum_x \sum_y \sum_i \sum_j T_{ijk}^{xy} n_j} \Big/ M \qquad [3.3]$$

The sum over x, y, j, and k (Equ. 3.4) gives the increase in mortality by months since last birthday R_i,

$$R_i = \frac{\sum_x \sum_y \sum_j \sum_k D_{ijk}^{xy}}{\sum_x \sum_y \sum_j \sum_k T_{ijk}^{xy} n_j} \Big/ M \qquad [3.4]$$

while the sum over x, y, i, and k (Equ. 3.5) results in *Rj*, the ratio of the deaths rates by current month as compared to average mortality.

$$R_j = \frac{\sum_x \sum_y \sum_i \sum_k D_{ijk}^{xy}}{\sum_x \sum_y \sum_i \sum_k T_{ijk}^{xy} n_j} \Big/ M \qquad [3.5]$$

Equ. 3.6 gives the age-specific death rates for those born in June (k=6) and December (k=12)

$$Q_{jk=6/k=12}^{x} = \frac{\sum_x \sum_y D_{ijk}^{xy}}{\sum_x \sum_y T_{ijk}^{xy} n_j} \qquad [3.6]$$

and the sum over all age groups x and time y results in the mortality trajectories over the time span of one year of life depending on the month of birth

$$R_{ijk} = \frac{\sum_x \sum_y D_{ijk}^{xy}}{\sum_x \sum_y T_{ijk}^{xy} n_j} \Big/ M \qquad [3.7]$$

The results of Equations 3.3 to 3.7 are displayed in Figures 3.1.A to 3.1.C. As expected, mortality increases by about 10% for males and females combined within twelve months (Fig. 3.1.A, solid lines females, dashed lines males). Mortality is lowest in August and highest in January; the difference is about 17.6% for males and 13.9% for females. The maximum difference in the death rates by month of birth is approximately 5.3% for males and 4.5% for females. The monthly pattern is similar for females

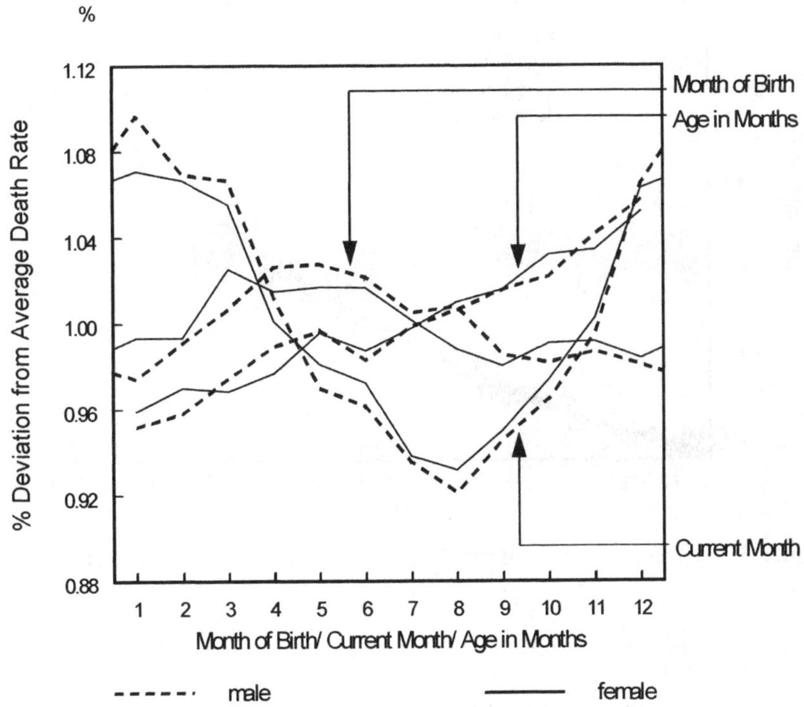

Figure 3.1.A Danish register data: deviations in standardized death rates by month of birth, current month and age in months.

and males. All three factors examined influence adult mortality simultane ously. This can be seen in Fig. 3.1.B and 3.1.C, where the mortality trajectories depend on month of birth, current month, and age in months. In Figure 3.1.B mortality differs considerably, depending on whether a person is born in June or December and therefore reaches a certain age during winter, when mortality is high, or during summer, when mortality is low. Figure 3.1.C shows the mortality trajectories over 12 months by month of birth. All three effects exist simultaneously and may interact with each other. In the following, an event-history model is used to estimate the force of mortality dependent on month of birth and corrected for interaction between month of birth and current half-year. The outcome of a model without the interaction term is described in Doblhammer (1999).

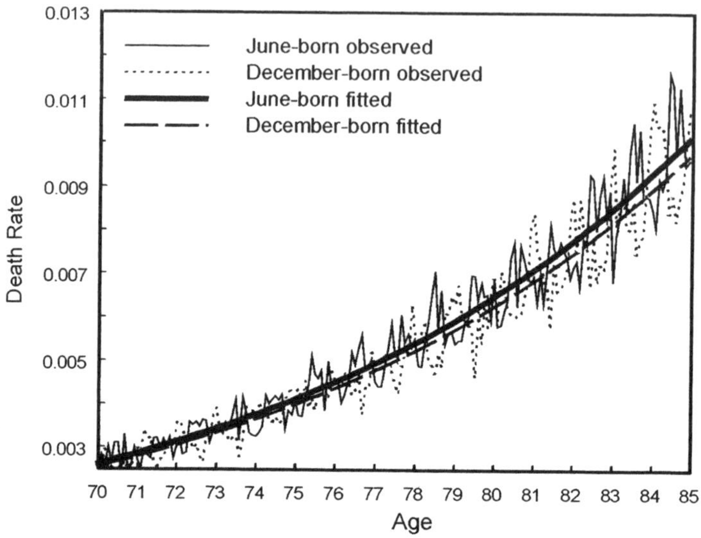

Figure 3.1.B Danish register data: death rates by age for June and December-born.

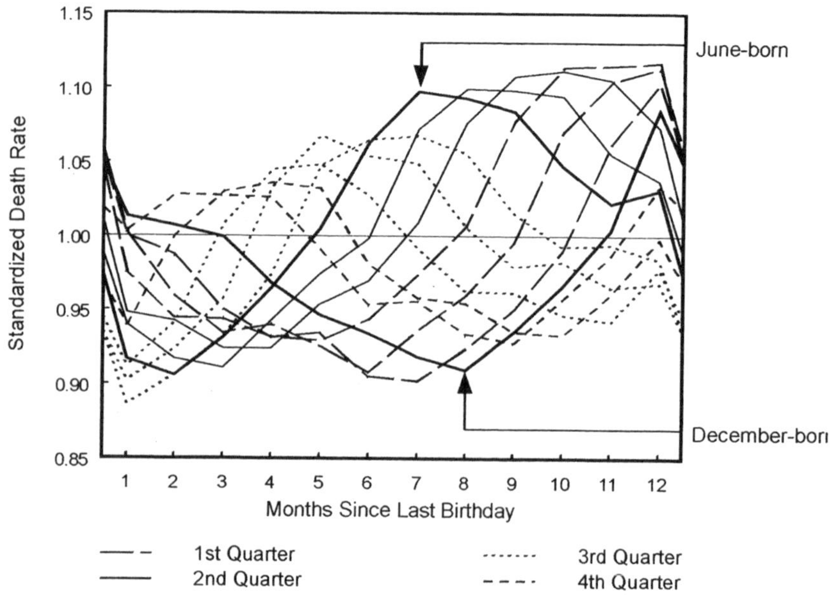

Figure 3.1.C Danish register data: death rates over the course of one year depending on the month of birth.

3.1.2 An Event-History Model to Separate the Effect of Month of Birth, Age, and Current Month

The longitudinal nature of the Danish dataset allows us to model $\mu(x)$, the force of mortality at age x, directly. The general mathematical specification of the model is

$$\mu(x) = \mu_0(x)\exp\left\{\sum_i \beta_i y_i(x)\right\} \qquad [3.8]$$

where $\mu(x)$ is a function of the baseline hazard $\mu_0(x)$; $y_i(x)$ denotes the value of the covariate and β_i the regression coefficient that measures the effect of the covariate *i* on the force of mortality. In this model the baseline hazard $\mu_0(x)$ is piecewise constant. The model assumes that the age-specific death rate for five-year age groups j follows a step function and that the death rate within the five-year age groups is constant.
Thus,

$$\mu_0(x) = a_j$$

with [3.9]

j = 50-54, 55-59, 60-64, 65-69, 70-74, 75-79, 80-84, 85-89, 90-95,95+

All other covariates are sets of binary variables that represent the levels of categorical covariates. The following covariates are included: sex; month of birth, which divides the year into thirteen periods, each consisting of four weeks. Week one starts at day one, week two at day eight, and so forth; current half-year, which is measured in two 6-months periods from January to June and July to December; and birth cohort: 1860-1879, 1880-1889, 1890-1899, 1900-1909, 1910-1918.

Month of birth and birth cohort are independent of time. The variable current half-year is time-varying and constructed such that in each half-year 50% of the average annual deaths occur. In the first half-year those with birth dates from January to June will be the oldest and will therefore experience increased mortality. In the second half-year the July to December-born will be the oldest and their mortality will be increased. Since in the first and second half-year the mortality risk is approximately the same the month of birth effect should disappear when an interaction effect be-

Table 3.1. Model 1: Relative mortality risks estimated by the hazard model defined in Equ. 3.8.

Variable	RMR	p-value	Variable	RMR	p-value
Month of birth		<.001	*Sex*		<.001
1.1–28.2.	1(RG)		Males	1(RG)	
29.1.–25.2.	1.007		Females	0.664	
26.2.–25.3.	1.031		*Cohort*		<.001
26.3.–22.4.	1.033		1860-1879	1(RG)	
23.4.–20.5.	1.034		1880-1889	0.969	
21.5.–17.6.	1.034		1890-1899	0.896	
18.6.–15.7.	1.026		1900-1909	0.842	
16.7.–12.8.	1.012		1910-1918	0.782	
13.8.–9.9.	1.009				
10.9.–7.10.	1.003		*Age*		<.001
8.10.–4.11.	0.999		50-54	1(RG)	
5.11.–2.12.	0.997		55-59	1.533	
3.12.–31.12.	0.996		60-64	2.373	
Current half-year			65-69	3.641	
1st half-year	1 (RG)		70-74	5.648	
2nd half-year	0.995		75-79	8.767	
			80-84	13.615	
			85-89	21.521	
			90-94	33.927	
			95+	52.818	

RG... reference group, RMR... relative mortality risk

tween month-of-birth and current half-year is constructed. The survival models were estimated with the program Rocanova.

Model 1 (Table 3.1) shows the main effects of the included variables without the interaction between month of birth and current month. The main effect of the month of birth reveals a 3 per cent excess in mortality of the March to July-born as compared to the January-born. Mortality decreases continuously over cohorts and is 22 per cent lower in the youngest cohort (1910-1918) than in the oldest cohort (1860-1879). The mortality of females is 64% of the mortality of males, the first and second half-year have equal mortality risks. The age-specific force of mortality approximately doubles with every 10-year age group.

Table 3.2. Model 2: Relative mortality risks estimated by the hazard model defined in Equ. 3.8. with an interaction term between month of birth and current half-year.

Variable	RMR	p-value	RMR	p-value
	Month of birth x current 1st half-year		Month of birth x current 2nd half-year	
Month of birth		<.001		<.001
1.1.–28.2.	1(RG)		1 (RG)	
29.1.–25.2.	1.025		0.993	
26.2.–25.3.	1.103		0.978	
26.3.–22.4.	1.149		0.952	
23.4.–20.5.	1.147		0.956	
21.5.–17.6.	1.120		0.974	
18.6.–15.7.	1.062		1.000	
16.7.–12.8.	1.012		0.011	
13.8.–9.9.	0.970		0.037	
10.9.–7.10.	0.962		0.031	
8.10.–4.11.	0.974		1.016	
5.11.–2.12.	0.965		1.019	
3.12.–31.12.	0.970		1.014	
Sex		<.001		
Males	1(RG)			
Females	0.644			
Cohort		<.001		
1860-1879	1(RG)			
1880-1889	0.968			
1890-1899	0.895			
1900-1909	0.842			
1910-1918	0.781			
Age		<.001		
50-54	1(RG)			
55-59	1.532			
60-64	2.370			
65-69	3.636			
70-74	5.640			
75-79	8.755			
80-84	13.598			
85-89	21.495			
90-94	33.890			
95+	52.765			

RG... reference group, RMR... relative mortality risk

Model 2 (Table 3.2) includes an interaction term between month of birth and current half-year. Taking again the January-born as the reference group one finds that in the first current-half year the February to August-born experience higher mortality and the September to December-born lower mortality. In the second half-year the picture is reversed. The August to December-born experience higher mortality while the January to July-born experience lower mortality. This is the expected pattern since in the first half-year those with birth dates January to July are the oldest, while in the second half-year those born between August and December are the oldest. The important result is that in the first half-year the excess mortality of those born in the first-half year is much higher (April and May-born: plus 15 % as compared to the January-born) than the excess mortality in the second half-year of those born in the second half-year (September and October-born: plus 4 % as compared to the January-born). Since approximately 50% of the deaths occur in each half-year the larger excess mortality of the spring-born cannot be the result of seasonal changes in the mortality risk but is rather caused by the month of birth.

In the following a third approach is presented that is based on death rates. The specification of the model varies from above hazard model. While in the hazard model an interaction term between month of birth and current half-year is used to explore the impact of the seasonal changing mortality risk on the month of birth effect, an interaction term between age since last birthday and current month is used in the next model.

3.1.3 A Model Based on Death Rates to Separate the Effect of Month of Birth, Age, and Current Month

For the Danish population it is possible to construct a surface of death rates by month of birth, current month and age in months using Equ. 3.1 and 3.2. The surface of the logarithm of the monthly deviations from the annual death rates is then modeled by the following Equ.:

$$\ln\left(R_{ij}^{xy}\right) = a_0 + \sum_{k=1}^{12} a_k I_{1k} + \sum_{k=1}^{12} b_k I_{2k} + \sum_{k=0}^{1} \alpha_k I_3 + \beta A + \sum_{k=1}^{12} \gamma_k A\, I_{2k} + u \qquad [3.10]$$

Table 3.3. The logarithm of the relative deviations from yearly Danish death rates by month of birth, age in months, current month, and sex.

Covariates	Parameter estimates	Covariates	Parameter estimates
Month of birth		*Current Month*	
January (RG)	0	January (RG)	0
February	0.008 *	February	-0.013
March	0.027 **	March	-0.022 **
April	0.032 **	April	-0.066 **
May	0.033 **	May	-0.098 **
June	0.035 **	June	-0.106 **
July	0.016 **	July	-0.131 **
August	0.014 **	August	-0.145 **
September	0.004 **	September	-0.126 **
October	0.003 **	October	-0.106 **
November	0.008 **	November	-0.079
December	-0.003	December	-0.021
Sex			
Male (RG)	0		
Female	-0.025 **		
Age	0.075 **		
Constant	-.047		

** $p < 0.01$, * $p < 0.05$, RG denotes the reference groups.

In Equ. 3.10 the variables I_1 to I_3 are indicator variables. They take the value one for a particular characteristic and zero otherwise. I_{1k} indicates month of birth (reference month: January), I_{2k} current month (reference month: January), I_3 sex (reference sex: males). The variable A stands for age since last birthday in months rounded up to the nearest integer. Two models are estimated. The first model only includes the main effects of the variables and results are displayed in Table 3.3. The second model includes the interaction term $\sum_{k=1}^{12} \gamma_k A\, I_{2k}$, which estimates the increase in the seasonal mortality risk with age (Table 3.4). The parameter values a_1, b_1, α_0 and γ_1 are set equal to zero; $u \sim N(0, \sigma^2)$.

Table 3.4. The logarithm of the relative deviations from yearly Danish death rates by month of birth, age in months, current month, sex and an interaction term between age in months and current month.

Covariates	Parameter estimates	Covariates	Parameter estimates
Month of birth		*Current month*	
January (RG)	0	January (RG)	0
February	0.009 **	February	0.006
March	0.027 **	March	-0.009 **
April	0.033 **	April	-0.043 **
May	0.033 **	May	-0.084 **
June	0.036 **	June	-0.089 **
July	0.018 **	July	-0.113 **
August	0.015 **	August	-0.130 **
September	0.005 **	September	-0.108 **
October	0.006 **	October	-0.081 **
November	0.011 **	November	-0.060 **
December	-0.001	December	-0.009 **
Sex		*Age*current month*	
Male (RG)	0	January (RG)	
Female	-0.025 **	February	-.0029
Age	0.010 **	March	-.0019
Constant	-.0645	April	-.0036
		May	-.0022
		June	-.0026
		July	-.0027
		August	-.0023
		September	-.0027
		October	-.0040
		November	-.0028
		December	-.0017

** $p < 0.01$, * $p < 0.05$, RG denotes the reference groups

As compared to previous two analyses of month of birth, age, and month of death (current month) the results only change slightly when the impact of the different factors is estimated based on the model described in Equation 3.3. Those born between April and June experience excess mortality of about 3% as compared to the January-born. Current mortality is about 15% lower in August than in January (Table 3.3). Introducing the interaction effect between current month and age does not change the parameter estimates for the month-of-birth effect (Table 3.4) and mortality of the April to June-born is still 3 % higher than of the January-born. This result again confirms that the differences in life span by month of birth are not caused by the seasonal fluctuations in mortality.

3.2 The Procreational Habits Hypothesis

The second hypothesis assumes that the causal mechanism is linked to socio-economic differences in the seasonal distribution of births. With only a few exceptions births are distributed seasonally over the year. It has been shown that both social and biological factors influence this seasonality (Lam & Miron 1996, Doblhammer et al. 2001). The hypothesis is that, if the seasonality in birth is partly driven by the preference of couples for certain seasons, then the intensity of the preference may differ between social groups.

In schizophrenia research this hypothesis is known as the procreational habits hypothesis. In their review Torrey at al. (1997) mention Huntington (1938) as the first to explain the seasonal birth pattern in individuals with schizophrenia by the idiosyncratic seasonal conception pattern of their parents. And they cite James (1978), who describes the exceptional conception pattern as

"in the summer, people wear fewer clothes in bed, and... a schizoid spouse is more likely then to notice his (or her) co-spouse there and accordingly to initiate sexual behaviour".

Other contemporary proponents of this theory are Hafner et al. (1987) and Bleuler (1991), who proposes that conceptions among depressed individuals are more likely to take place in spring, when depression is less severe. In their review Torrey et al. come to the conclusion that the procreational habits theory can be ruled out as an explanation for the winter-spring excess births among schizophrenics. First, the theory predicts that individuals with schizophrenia who have a family history of the disease should have a greater winter-spring birth excess, which is clearly not the case. Second, the procreational habits theory also predicts that siblings of individuals with schizophrenia should also have a winter-spring birth excess. Torrey et al. (1997) cite six studies that did not find a winter-spring birth excess among siblings of people with schizophrenia.

To test this hypothesis in the context of mortality, education was used as an indicator of social group. Parental education is not contained on birth certificates from the beginning of the 20th century. It was thus assumed that, in 1991, a person's educational level was closely linked to the educational level of his or her parents. The Austrian microcensus of the year 1991 reveals that the educational status of the 15- to 19-year-old Austrians depends to a large extent on the social and educational status of the parents (Schwarz et al. 2002a, Schwarz et al. 2002b), despite tuition-free access to all levels of education since the 1970s. For earlier birth cohorts that did not benefit from the expansion of the Austrian education system, the intergenerational correlation in education must have been even stronger than it is today.

We calculated the seasonal distribution of the birth dates of Austrians aged 50+ by educational group on the basis of the 1981 census. It appears that the spring peak in births is stronger among adults with high or medium education, while people with basic education are over-represented among the autumn-born (Fig. 3.2).

The correlation between mean age at death by month of birth and the deviations in the monthly birth distribution from the average monthly pattern is -0.78 (Pearson correlation, one-sided test: $p<0.001$) for adults with high education, -0.84 ($p<=0.001$) for adults with medium education, and 0.88 ($p<0.001$) for adults with basic education. A recent study on the seasonality of births in the former Czechoslovakia comes to the same results (Bobak & Gjonca 2001). A similar result was found on the basis of a 10% sample of the 1971 census of economically active, British-born males. Non-manual workers tended to be born in spring and manual workers in autumn and winter (Smithers et al. 1984).

One can therefore conclude that the month-of-birth pattern in longevity is not due to differences in the conception pattern by socio-economic group. On the contrary, it appears that the differences in life span would be even larger if socio-economic differences were taken into account.

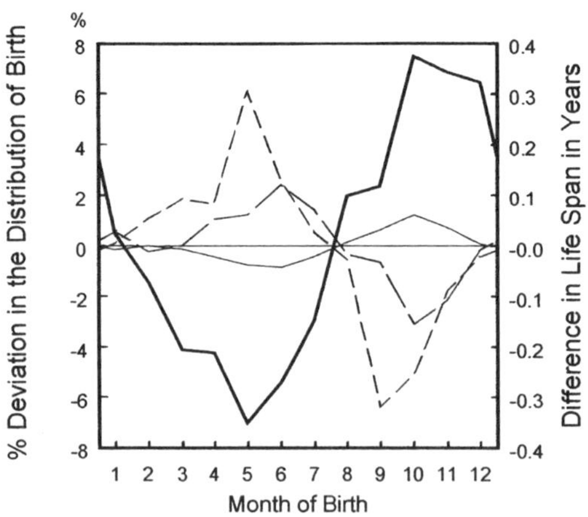

Figure 3.2. Per cent deviations in the seasonal distribution of births of Austrians (ages 50+) by education from the average seasonal distribution of births: medium (small dashed line), high (long dashed line), basic education (solid line) and differences in mean age at death (bold solid line).

3.3 The Deadline Hypothesis

Another prominent hypothesis in sociological and psychological research is that children who are born after a certain deadline have to wait another year before they can enroll at school, which makes them about a year older than the youngest children in their school class. For example, at the beginning of the 20th century school started on 1 October in Austria. Those children who had not turned six before this date had to wait another year. Thus, children born in the autumn experienced an age advantage over their classmates.

Research suggests that this age advantage affects scholastic aptitude, which might translate into a lifelong advantage in various regards. Alton and Massey (1998) studied the General Certificate of Secondary Education (GCSE) examinations. Almost all pupils in England, Wales, and Northern Ireland take this examination at age 16. Starting with pupils born in September and moving to those born in August, i.e. proceeding from the oldest to the youngest pupils of a given school year, the authors find that there is a steady decline in marks. By the age of 18 these differences have vanished. The authors attribute this to a selection effect stemming from the fact that only those students who did well at the GCSE examinations will have continued to the advanced level. A significant interaction between reading failure and age category (overage at school entrance vs. correct age) by season of birth was observed amongst second-year pupils (Flynn et al. 1996). Although the authors mention entrance cutoff birth dates to kindergarten as a probably explanation they do not rule out that the reported associations may be attributed to selective sampling.

The effect of relative age has been widely tested in the area of sports. Dudink (1994) finds a significant relationship between date of birth and success in tennis and soccer. In both disciplines half of the tennis players in the top rankings of the Dutch youth league and a third of the English soccer players in league clubs were born during the first three months of the competition year. The consistent asymmetry in the birth-distribution of senior professional soccer players has led Helsen et al. (1998) to investigate whether such asymmetries already exist in youth categories in soccer. Their results indicate that young players born from August to October are more likely to be identified as talented at ages 6 to 8 and are exposed to higher levels of coaching. These players are born early in the selection year and are relatively older. Those born relatively late in the selection year tend to dropout as early as 12. A study of the effect of a horse's month of birth on its future performance in sport (Langlois & Bloin 1998) showed that, under the age of four, horses that were born in the second part of the year were particularly disadvantaged in flat races. The disadvantage was not so strong for jumping races and for trotters. After the age of four the effect was of much smaller magnitude. They suggest that two effects may be responsible for the results: a relative age-effect, which is more prominent in races where older horses are favoured. This relative-age effect decreases with age (probably due to selection). The second effect is seen mainly among older horses in equestrian competitions and is not further specified by the authors.

Angrist and Krueger (1991) performed the most widely known study exploring the deadline effect with regard to outcomes in schooling and earnings. However, the authors argue the other way round from the studies cited above. Since compulsory schooling laws require students to stay in school until their 16th or 17th birthday, the oldest students may drop out slightly earlier, which should be reflected in their later earnings. In their study they exploit an "unusual natural experiment" to estimate the effect of compulsory schooling laws on educational attainment and earnings. They explain the experiment as follows:

"The experiment stems from the fact that children born in different months of the year start school at different ages, while compulsory schooling laws generally require them to remain in school until their sixteenth or seventeenth birthday. In effect, the interaction of school-entry requirements and compulsory schooling laws compel students born in certain months to attend school longer than students born in other months. Because one's birthday is unlikely to be correlated with personal attributes other than age at school entry, season of birth generates exogenous variation in education that can be used to estimate the impact of compulsory schooling on education and earnings" (p. 980).

In their study they find that children born in the first quarter of the year have a slightly lower average level of education than children born later in the year. According to the authors, school districts typically require a pupil to have turned six by January 1 of the year in which he or she enters school. Therefore, those born at the beginning of the year enter school at an older age and attain the legal dropout age earlier in their educational careers than people born later in the year. In support of their study they report that the seasonal pattern of education is not evident in college graduation rates or in post-graduate degree completion rates. Since compulsory schooling laws do not affect these two types of institutions, the authors conclude that the relationship between years of schooling and month of birth is entirely due to compulsory schooling laws. In the second part of their paper they showed that not only educational attainment but also earnings later in life are related to the month of birth.

In a recent paper by Pflug (2001) the author repeats the study by Angrist and Krueger for the Netherlands and presents evidence that the season of birth influences school performance not in terms of compulsory schooling effects but in the area of learning abilities. They refer to the arguments presented above – that it is the relative age among class mates that influences learning ability and, as a result of education, earnings later in life.

Further criticism of Angrist and Krueger's study was raised by Bound and Jaeger (1996). They showed that the association between quarter of birth and earnings or other labour-market outcomes existed for cohorts that were not bound by compulsory school attendance laws. Furthermore, Angrist and Krueger did not take into account the fact that differences in the birth distribution of social groups exist and that other factors, such as health, may be related to the season of birth.

Based on the studies discussed above, there seems to exist firm evidence that the relative age among pupils influences their scholastic aptitude and their sporting success. In Austria at the beginning of the 20th century school started on 1 October and children who had not turned six before this date had to wait another year. Thus, the autumn-born were the oldest among their classmates. Nevertheless one finds that the autumn-born achieve lower educational levels than the spring-born, which is contrary to the evidence presented above. In addition, at the turn of the 20th century in Austria, many schools, particularly in rural areas, combined several years into one class and a single teacher taught the different levels.

3.4 Selective Survival or Debilitation

The fourth hypothesis assumes that selective survival during infancy is the causal mechanism that explains the relationship between month of birth and life expectancy. Specifically, the hypothesis is that autumn-born infants suffer higher mortality in their first year of life than spring-born infants. This would leave the relatively more robust individuals alive, who would experience lower mortality at adult ages.

Infant mortality decreases dramatically over the first year of life. Looking at births and infant deaths in Denmark between 1911 and 1915, the probability of death decreases from 0.023 in the first 24 hours of life to 0.00275 in the twelfth month of life. At the same time, infant mortality is highly seasonal. However, the pattern of seasonality depends on age. Although at almost all ages death rates are highest in the winter months in the

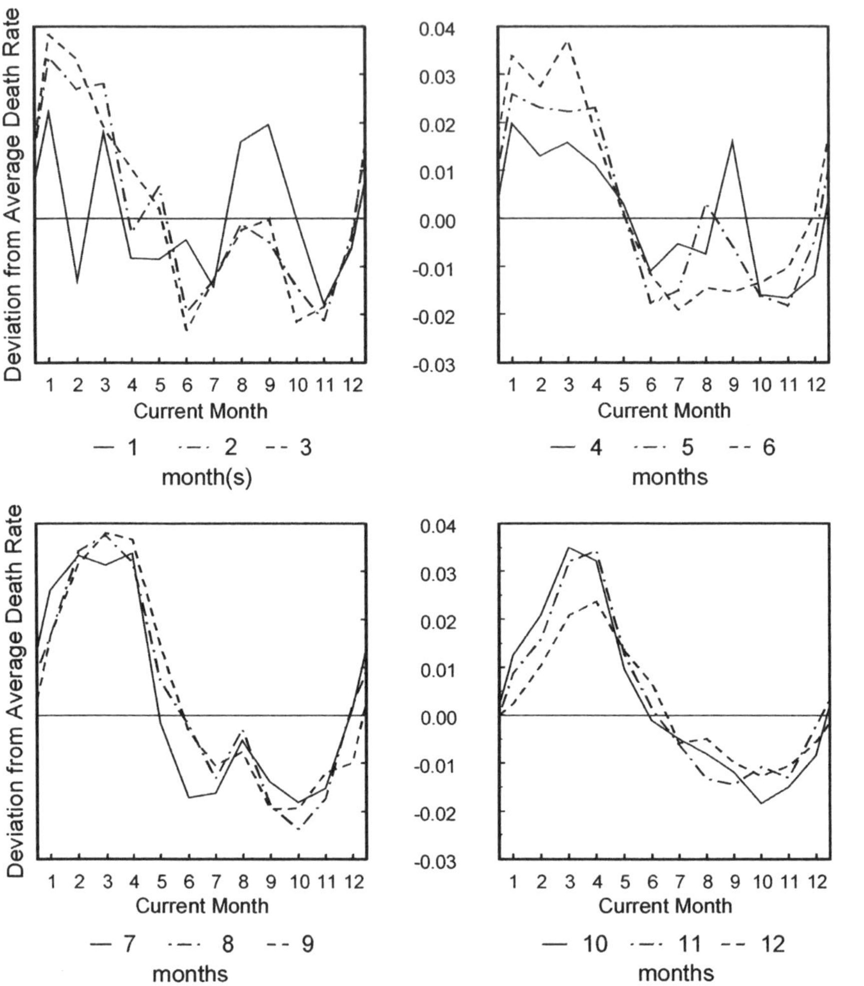

Figure 3.3. Deviation from the average death rate at a particular age (in months) by month of the year (current month) in Denmark 1911–1915.

first five months of life, a prominent secondary peak exists for the summer months of July, August, and September. This secondary peak tends to become smaller and even to disappear from the sixth month of life onwards (Fig. 3.3). In a model of the effect of the month of birth on survival during the first year of life one therefore has to account for the changes in the seasonality of deaths with age and for the strong decrease in the death rate over the first twelve months.

To test whether selective survival or debilitation during the first year of life explains the differences in remaining life expectancy at age 50, data on infant mortality for Denmark aggregated over the five-year period 1911-1915 were used. The tabulations in the "Statistik tabelvaerk " provide the number of deaths by age of the infant in months and month of death. Thus, the actual date of birth of the deceased infants can fall within a period of two months – on average it falls on the first day of each month. On the basis of the monthly number of births aggregated over the period 1911–1915, death rates for each month of the first year of life were then calculated. In order to center the number of births at the first day of each month the average number of births of two adjacent months are used in the calculations.

To estimate the impact of month of birth on the monthly death rates in the first year of life a model was used that is similar to that employed for adult ages. The logarithm of the monthly deviations from the annual death rate was modeled by applying equation [3.11]:

$$\ln\left(R_{ij}\right) = a_0 + \sum_{k=1}^{12} a_k I_{1k} + \sum_{k=1}^{12} b_k I_{2k} + c_1 \ln x + \sum_{k=1}^{12} \alpha_k \ln x I_{2k} + u \qquad [3.11]$$

The variables I_1 and I_2 are indicator variables. They take the value one for a particular characteristic and zero otherwise. I_{1k} indicates month of birth (reference month: January), I_{2k} current month (reference month: January). Age x is in months rounded up to the nearest integer. To reduce the number of parameters in the model and because the decline in mortality during the first year of life can be modeled by a logarithmic function, x is included as a numerical variable. The term $\sum_{k=1}^{12} \alpha_k \ln x \, I_{2k}$ corrects for the interaction in the death rates between age and current month. The infant mortality surface consists of only 144 data points (12 birth months x 12 ages). Monthly births are only available for both sexes combined and infant deaths by age in months and month of death are aggregated over the five-year period 1911-1915. Hence, the model does not include sex and cohort effects. The parameter values a_0, a_k, b_k, c_k and α_k were estimated by least squares regression; $u \sim N(0,\sigma^2)$.

When applied to Danish data on seasonal infant mortality in the first year of life from the years 1911 to 1915, the model in Equ. 3.11 seems to indicate that infants born in June are the most vulnerable. Table 3.5 contains the parameter estimates of the model described in Equ. 3.11 and those of a model without correction for the interaction between month of death and age. Both models fit the data well, with an adjusted R^2 of 0.93 for the model without correction of age at death and current month and an adjusted R^2 of 0.96 for the model with correction. The latter model predicts that the standardised death rate during the first year of life is about 31% higher for the June-born than for the February-born. The correction for the interaction between the seasonality of death and age has two effects. First, it decreases the mortality disadvantage of the June-born and, second, it shifts the whole pattern to the left.

This is not the first study to recognize that infant mortality depended strongly on the season of birth historically and that the seasonal pattern of infant deaths depends on age. Already in 1938 Huntington presented similar results for Belgium and New York (Huntington 1938). Breschi & Bacci

Table 3.5. Effect of the seasonality of death on infant mortality by month of birth.

	Model 1		Model 2	
	Parameter estimate	p-value	Parameter estimate	p-value
Intercept	-1.529	0.0000	-1.352	0.0000
Current month				
January	0.358	0.0000	0.241	0.1462
February	0.421	0.0000	0.084	0.6175
March	0.506	0.0000	0.050	0.7673
April	0.451	0.0000	-0.165	0.3345
May	0.217	0.0052	-0.147	0.3916
June	-0.017	0.8188	-0.339	0.0498
July	-0.122	0.1122	-0.241	0.1616
August	-0.052	0.4955	-0.068	0.6898
September	-0.174	0.0241	0.072	0.6715
October	-0.291	0.0002	-0.033	0.8433
November	-0.232	0.0028	-0.060	0.7164
December	0.000		0.000	
Month of birth				
January	-0.072	0.3488	-0.044	0.5058
February	-0.070	0.3572	-0.023	0.7308
March	-0.023	0.7672	0.018	0.7964
April	0.080	0.2983	0.095	0.1662
May	0.206	0.0078	0.165	0.0179
June	0.304	0.0001	0.233	0.0010
July	0.301	0.0001	0.195	0.0054
August	0.254	0.0011	0.136	0.0486
September	0.203	0.0088	0.081	0.2334
October	0.112	0.1436	0.025	0.7099
November	0.020	0.7904	-0.023	0.7328
December	0.000		0.000	
ln (age)	-0.854	0.0000	-0.938	0.0000

Table 3.5. (continued)

	Model 1		Model 2	
	Parameter estimate	p-value	Parameter estimate	p-value
*Interaction ln(age)*current month*				
January* ln(age)			0.071	0.4390
February*ln(age)			0.203	0.0305
March*ln(age)			0.274	0.0042
April*ln(age)			0.369	0.0002
May* ln(age)			0.218	0.0234
June * ln(age)			0.193	0.0448
July * ln(age)			0.071	0.4549
August * ln(age)			0.010	0.9197
September * ln(age)			-0.148	0.1179
October * ln(age)			-0.155	0.0970
November * ln(age)			-0.103	0.2574
December * ln(age)			0.000	

(1996) came to a similar conclusion for Switzerland and Belgium, as did Lumaa et al. (1998) for Finland and Reher & Sanz-Gimeno for Spain (2003).

If selective survival during infancy explains the differences in adult life span, then one would expect a significant negative correlation between the probability of death in the first year of life and after age 50. If debilitation *in-utero* or during the first year of life (our fifth hypothesis) is the causal mechanism, then one would expect a significant positive correlation because rates of debilitation are highly correlated with mortality.

For Denmark the correlation between infant mortality in the first year of life and adult mortality after age 50 (Figure 3.4) is 0.87 and statistically significant ($p<0.0001$). This result clearly rules out the selection-hypothesis and indicates that the causal mechanism of the month-of-birth pattern in life span is related to debilitating factors that affect either the mother during pregnancy or the infant in the first year of life.

3.5 Conclusion

There are four main conclusions that can be drawn from the analysis of data from Austria, Denmark, Australia, and Hawaii:

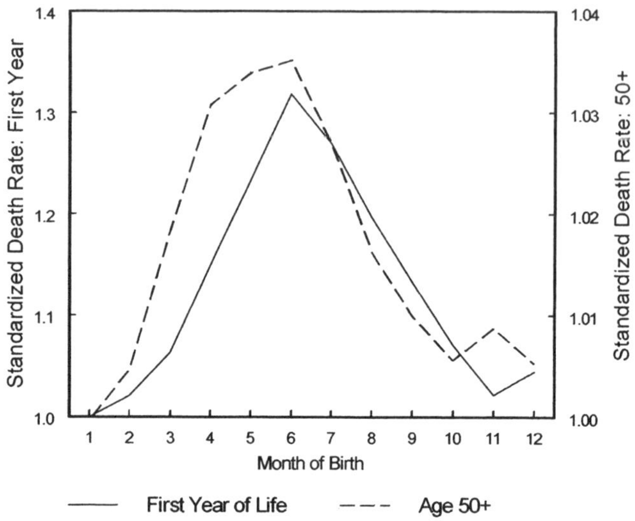

Figure 3.4. Standardized death rates of Danish infants born in a specific month and difference in remaining life expectancy at age 50 by month of birth.

First, the differences in life span by month of birth are tied to the seasons of the year. This is clearly shown by the reversal of the pattern in the Northern and the Southern Hemispheres.

Second, the pattern also exists in regions close to the equator: Hawaii is one example, Queensland in Australia is another. This finding rules out the explanation that differences in the length of daylight are the underlying causal mechanism. This hypothesis will be further explored on the basis of the US death data (Chapter 5).

Third, although Queensland and Hawaii are close to the equator these two countries experience two distinct seasons of the year, namely a wet and a dry season. In both countries those who experience a mortality advantage later in life are born at the beginning of the season when temperatures are more moderate. In non-tropical regions with four distinct seasons those born in autumn and winter have longer lives.

Fourth, within a country the pattern may vary greatly among different groups. This is shown by the analysis of the month-of-birth pattern for Aborigines in Australia. Not only is the peak-to-trough difference for them more than nine times as large as for other native-born Australians – the pattern is also different. The next chapter will explore whether this is also true for different social and racial groups in the United States.

The test of the five hypotheses clearly reveals that the month-of-birth pattern is not the result of an age-artifact caused by the seasonality of death, nor is it the result of social factors such as social differences in the seasonal distribution of births or age at entry into school. There is no support for the hypothesis that differential infant survival is the causal factor behind the observed phenomenon. On the contrary, for Denmark there exists a strong and highly significant positive correlation between the differences in infant mortality and the differences in mortality after age 50 by month of birth, which is consistent with the hypothesis that debilitation early in life is the causal mechanism.

The results support the theory of there being a critical period early in life that affects adult health and survival without other mediating life-course factors. The model of infant mortality does not permit any conclusions about whether debilitation occurred *in-utero* or during early infancy. The results are consistent with both the effect of seasonal differences in nutrition and of seasonal differences in the disease environment early in life. In past decades the food supply in general, and the availability of fresh fruit and vegetables in particular, differed from season to season. Mothers who gave birth in autumn and early winter had access to plentiful food and fresh fruit and vegetables throughout the third trimester of pregnancy, the period of peak growth of the fetus; those who gave birth in spring and early summer experienced longer periods of inadequate nutrition during their last trimester. On the other hand, infant mortality was historically primarily the result of infectious disease, which occurred seasonally. The close correlation between infant mortality by month of birth and the month-of-birth pattern in adult life span therefore suggests that infectious disease plays an important role.

In the Danish data, differences in adult life span by month of birth are significantly smaller in the more recent cohorts than in the oldest cohorts. In a previous study by Doblhammer & Vaupel (2000) it is shown that among those born before 1888 the differences are amplified by a factor of 72%. That is, the difference in adult mortality by month of birth is almost twice as large for the cohorts born from 1863 to 1888 as it is for those born between 1889 and 1918. Chapter 6 in this book provides a more detailed analysis of the cohort changes in the month-of-birth pattern. This shift lends further support to the explanation that the relationship between month of birth and adult life expectancy is caused by prenatal and early postnatal conditions related to nutrition and infectious disease, since considerable improvements in maternal and infant health took place between these two periods (Preston & Haines 1991).

4 Differences in Life Span by Month of Birth in the United States

The most important criticism about research on early-life environment and later-life health and mortality is concerned with how this research is interpreted (for an overview see Kuh & Davey Smith 1993). It has been argued that the strong correlation between early environment and adult mortality may simply be an effect of continued deprivation over the whole life course rather than of factors that act early in life (Ben-Shlomo & Davey Smith 1991).

This study uses month of birth as an indicator for early-life circumstances. The previous chapter provided evidence that the month of birth does not have any further life-course interpretations that affect mortality. The comparison of the pattern for people born in the Northern Hemisphere to those born in the Southern Hemisphere suggests that the effect of the month of birth on the life span is related to the seasons of the year. In both hemispheres, the remaining life expectancy after age 50 is lower for people born in the spring than for those born in the autumn. Thus, in the Northern Hemisphere the pattern is a mirror image of the Southern Hemisphere's pattern.

Furthermore, the previous chapter showed that a significant positive correlation exists between infant mortality in the first year of life by month of birth and mean age at death by month of birth. This led to the conclusion that it is debilitation rather than selection that leads to the differences in life span by month of birth.

The aim of this chapter is to gain deeper insights into the mechanisms that cause the differences in life span by month of birth by using information about the social, demographic, and geographic characteristics of the decedents. It explores whether the differences vary with educational background, marital status, region of birth, race, and sex on the basis of 15.6 million US death certificates for the years 1989 to 1997.

One would expect that the differences depend on social factors: they should increase with lower education and they should be larger for African-Americans than for whites. Education is largely heritable between generations and low education can thus be considered as an indicator for poor living circumstances of the decedent's parents and therefore also of the decedent's first period of life. Negative seasonal factors such as insufficient nutrition or infectious disease are more common and also more harmful under poor living circumstances than under more prosperous ones. The same argument holds for African-Americans and whites.

Evidence exists that marital status is partly determined by health selection into marriage (Goldman, Lord & Hu 1993) and that the never-married are frailer than the ever-married population (Goldman & Hu 1993). Differences in life span by month of birth should therefore be larger among the never-married than among the ever-married.

The information about state of birth allows to test four hypotheses that have been put forward to explain the month-of-birth effect. First, the state of birth provides information about the approximate latitude of birth and thus the seasonal differences in the hours of daylight. It has been shown that the seasonal changes in the hours of daylight influence the human neuroendocrine functions (Wehr 1998). Several researchers have postulated that the winter-spring birth excess observed in schizophrenia and bi-polar disorders might be caused by variations in internal chemistry or neural development brought about by seasonal variations in light (Turnquist 1993, Quested 1991, Morgan 1978, Jongbloet 1975, Pallast et al. 1994). If the effect of month of birth on mortality later in life is due to seasonal differences in daylight, then this effect should be correlated with latitude and should be smaller in the southern part of the US than in the northern part.

Second, the state of birth provides information about the prevailing macro climate. According to the widely used Köppen classification (Köppen 1936), the United States consists of six major climatic zones. Regions between 30 and 60 degrees North latitude belong to the *humid continental climate* zone characterized by changeable weather, hot summers and cold winters. On the eastern side of the continent between 20 and 35 degrees North latitude, the *humid subtropical climate* prevails. This climate is characterized by uniform precipitation throughout the year with moist and hot summers and mild winters. The *mild-latitude steppe* and *desert climate* are found in the continental interior and include the tropical and subtropical steppe climates in the South. Temperature conditions are extremely variable, with the annual means decreasing and the annual range increasing to the North. In the higher latitudes, winters are severely cold with meager precipitation. In the South the tropical and subtropical steppe climate is located primarily on the periphery of true deserts. The *Mediterranean Climate* of Southern California features hot dry summers and cool wet winters. The *Marine west coast climate,* which extends between 35 and 60 degrees North latitude, is characterised by ample precipitation in all months, small annual temperature ranges, mild winters and moderate summers with rarely more than 20 degrees Celsius. This area is confined to a narrow coastal strip of the West Coast.

The magnitude of seasonal differences in temperature depend on the climate. Huntington (1938) was one of the first to formulate the hypothesis that temperature, in particular high temperature at the time of conception, causes the differences in life span by weakening the "germ plasma".

"…..season of birth is related to length of life primarily, although perhaps not wholly, through the condition of the parents and hence of the germ plasma at the time of conception. This by no means excludes the hypothesis of a permanent depressive effect when a very young infant experiences undue heat or disease regardless of any selection through death". (p. 186)

More recent research has shown that the sperm quality of men who work outdoors decreases during periods of high temperature (Centola & Eberly 1999). Pasamanick (1986) proposed that hot summers were the cause of protein deficiency at the time of conception. As temperatures are generally highest in July/August, infants born in March and April should experience the lowest life expectancy. The peak-to-trough difference should be largest in regions with high summer temperatures.

The third hypothesis is that infectious disease causes the differences in life span by month of birth. The information about state of birth is linked to the burden of infectious diseases imposed by the climate. Climate influences the incidence of infectious diseases through temperature, humidity, and precipitation in the form of normal and severe rainfall, wind and ocean currents. Thus, the incidence of infectious disease is generally highly seasonal. *Vectorborne diseases* such as malaria and dengue fever were once common in the southern United States. The incidence of waterborne diseases, which account for as much as 80 percent of the annual mortality due to infectious diseases (Clark 1993), is correlated with warmer temperatures and flooding (Colwell 1996, Gueri, Gonzalea & Morin 1986). The transmission of infections via food in the form of viruses, bacteria, and protozoa causes significant illness. Campylobacteriosis is a common diarrhoeal disease of bacterial origin, which is generally acquired by drinking contaminated milk or water or by eating improperly cooked meat or fowl. There is a pronounced seasonality in its incidence since it normally occurs during the spring (Korlath et al. 1985). The incidence of other foodborne diseases, such as salmonella enteritidis (Shears & Wright 1985) and cyclospora cayetanenis (Madico et al. 1997), coincides with peak climatic temperatures. The seasonal variability of many respiratory diseases suggests the existence of a weather-related factor in the transmission of airborne diseases. The respiratory syncytial virus, which peaks in the autumn (Shears & Wright 1995), commonly causes bronchitis and mild upper-respiratory-tract infections that can result in severe, and even fatal lower tract problems, especially among infants.

The southern part of the US should have been an unhealthier environment for infants (when controlled for urban/rural residence, race and socioeconomic status) because of a higher incidence of infectious and parasitic disease and gastroenteritis. An increase in the peak-to-trough difference from North to South would thus suggest that infectious disease during childhood plays a major role in old-age mortality.

On the other hand, respiratory diseases, which are mainly caused early in life by seasonal virus infections, were spread more in the poorly ventilated conditions accompanying winters in the North[1] (Preston & Haines 1991).

The fourth hypothesis is that the nutrition of the mother during pregnancy and of the baby in the first year of life is the causal mechanism that underlies the month-of-birth pattern. Diet is the result of a complex interaction of social, economic, and environmental factors. At the beginning of the 20th century, large regional differences in diet existed in the US. What the different diets had in common was that they do not resemble contemporary US dietary patterns (Levenstein 1988, 1993). People ate meat – either fresh meat in the middle or upper classes, or salted meat among the poor – with potatoes or some other starchy staple. Other vegetables and fruits occupied only a small niche on the table. The first vitamin was only discovered in 1911; in the early 1900s nutritionists were actually even opposed to greens, which were thought to require more bodily energy for digestion than they provided (Dyson 2000). With the notable exception of the South, there is no indication that even the poorest workers suffered from insufficient quantities of food. The problem was rather the quality and the variety, especially during the winter and spring, when affordable supplies dried up. As mentioned above, there were large regional differences in diet. Diet was particularly poor in the rural South, where farmers – many of whom were tenant farmers – survived on the "three m's", that is, meat (salted pork), meal (mainly corn meal) and molasses. Sharecroppers moved too frequently to raise their own vegetables or livestock. As a result, milk and fresh meat, even fresh pork, were usually rarities on croppers' tables. The tenancy system discouraged the production of non-cash crops and tenants depended on the sale of flour, molasses, sugar and salt pork, which mainly came from the large meat factories in Chicago. In the

---

[1] The text refers to New England, the Middle Atlantic, East North Central, and West North Central as the North; to the South Atlantic, East South Central and West South Central as the South; to the Mountain Region and Pacific as the West.

summertime, greens and berries could be gathered, but the hard work required during planting time in the spring and harvest time from August to December normally inhibited those endeavours, particularly in the cotton belt (Levenstein 1993). Contemporary dietary studies have shown that, particularly in winter and early spring, diets in the rural South lacked variety whereas better nutrition and a wider variety of foods was shown to exist in the urban South (Dirks & Duran 2001).

In summary, in the North and in urban areas people suffered from seasonal deficiencies in nutrients while in the rural South they experienced deficient nutrition during the whole year, with downright malnutrition in winter and early spring.

The effects of region, birth, race, and education are inherently confounded. There are two other confounding factors, namely, age and cohort. Chapter 6 will show that the differences in life span by month of birth are smaller in more recent cohorts and that they become smaller with age. Due to the cross-sectional nature of the US death records one cannot distinguish between age and cohort effects. However, the calculations are age-standardised when the data are analysed by univariate methods, and they control for the main effects of sex, education, race, and region on age at death used when multivariate methods are applied.

4.1 Data

This chapter is based on the publicly available Multiple Cause of Death Data for the years 1989 to 1997 from the National Center for Health Statistics (NCHS), to which the information about month of birth was added. This study consists of 15,596,952 death records for all black and white US-born decedents who died in the United States between 1989 and 1997. Table 4.1 gives an overview of the data regarding mean age at death and the proportion of the decedents that fall into the different categories of education, marriage status, race, and region of birth. Two educational groups are distinguished: the 5.3 million decedents with a basic education have less than a high school education or did not finish high school (up to eleven years of education); the 8.2 million decedents with a high education have at least finished high school. There is also a residual group of 2.2 million death records with an unknown educational level. The analysis is restricted to blacks and whites to guarantee that a sufficient number of deaths were given for each month of birth. The states of birth are grouped

Table 4.1. Number, proportion, and mean age at death of US decedents by region, sex, education, marital status, and race.

	Total	Mean age at death	%
Region of birth			
New England	928,642	77.10	0.06
Middle Atlantic	2,950,756	76.83	0.19
East North Central	2,900,962	77.05	0.19
West North Central	1,855,131	78.68	0.12
South Atlantic	2,409,504	75.34	0.15
East South Central	1,666,643	75.93	0.11
West South Central	1,839,805	76.12	0.12
Mountain	448,687	76.12	0.03
Pacific	579,369	74.87	0.04
Sex			
Males	7,665,738	74.27	0.49
Females	7,931,214	78.86	0.51
Race			
Whites	13,871,496	77.03	0.89
Blacks	1,725,456	73.20	0.11
Education			
Low (<12 years)	5,261,840	78.35	0.34
High (>=12 years)	8,164,552	75.43	0.52
Unknown	2,170,560	76.80	0.14
Marrital status			
Never married	1,046,611	75.15	0.07
Married	6,844,464	72.77	0.44
Widowed	6,367,655	82.42	0.41
Divorced	1,305,106	69.60	0.08
Unknown	33,116	72.80	0.00
TOTAL	15,596,952	76.60	100.00

according to the grouping in the death records into the nine regions of New England, Middle Atlantic, East North Central, West North Central, South Atlantic, East South Central, West South Central, Mountain and Pacific; Hawaii and Alaska are excluded.

4.2 Methods

All univariate results are age-standardised. Let $p_j(x)$ be the fraction at age x of deaths in group j and $p(x)$ be the respective proportion in the standard population, which is the total black and white US population between ages 50 and 100. Each age x in group j is re-weighted by $W_i(x)=\frac{p(x)}{p_j(x)}$, such that for each group j the same mean age at death is derived.

Equation 4.1 estimates the simultaneous effects of the characteristics of the decedents on their age at death

$$x_i = \alpha_0 + \beta' Y_i + \gamma' I_i + u_i . \qquad [4.1]$$

Let x_i be the age at death, Y_i the matrix of the indicator variables month of birth, sex, education, race, region and marital status, and I_i the matrix of the two-way interactions between the variables; α_0, β', and γ' are the parameter estimates. The error term u_j follows a normal distribution with mean zero and variance σ^2. The above model assumes that the observed data stem from normal distributions with equal variances but with different means. The descriptive analyses show that the distribution of ages at death deviates significantly from a normal distribution and that the variances of the different categories are not homogenous. The differences in the variances are small, however, and mainly significant because of the large numbers of observations. The sensitivity of the model to the violation of the model assumptions was tested by using different starting ages, and by using log and rank transformations of age. It appears that the results were largely insensitive to these changes.

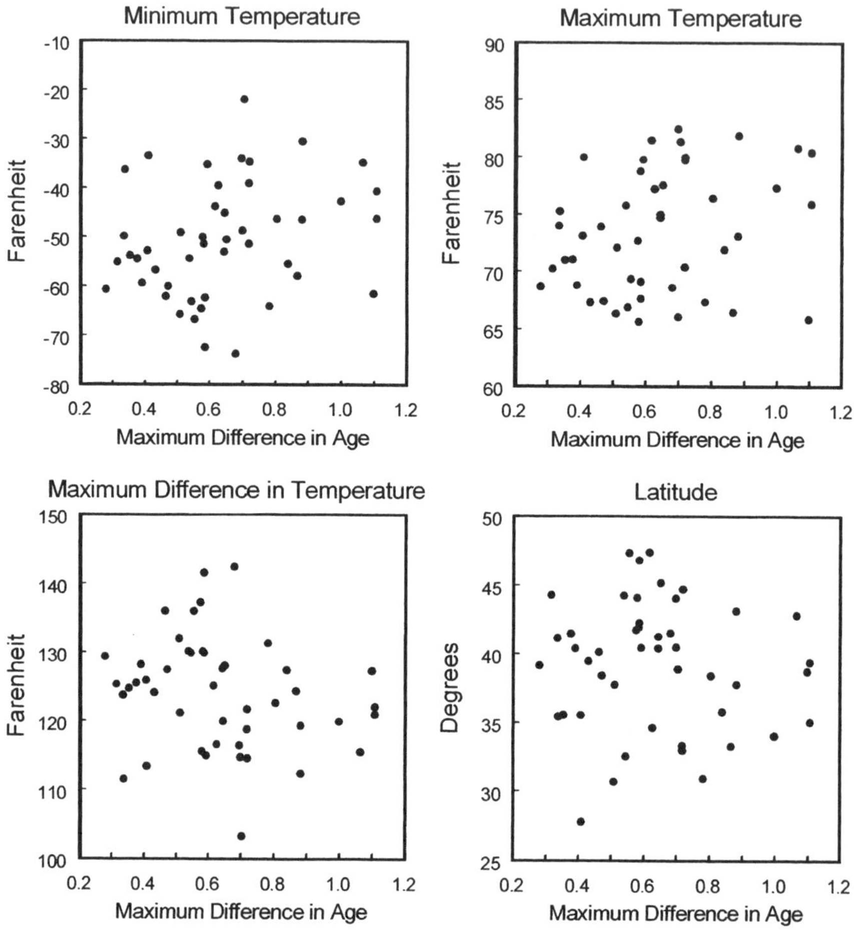

Figure 4.1. Scatterplots of the maximum difference in mean age at death and maximum/minimum temperature, the maximum difference in temperature, and latitude for the 48 contiguous US states.

4.3 Results

A highly significant difference in mean age at death by month of birth exists for US decedents who died between 1989 and 1997. Those born in June and July die about 0.44 years earlier than the October-born. The difference is slightly larger for males (0.434 years) than for females (0.402 years).

Latitude and Temperature

There is no positive correlation between the peak-to-trough differences in the mean ages at death and the latitudes. On the contrary, a small borderline significant negative correlation (Pearson correlation ρ=-.27, p=.06) exists. The correlation of the month-of-birth effect with the maximum difference in temperature, another indicator for latitude, is also negative but not significant (ρ=-.20, p=.18). There is a small positive correlation between the peak-to-trough difference in mean age at death and both maximum (ρ=.26, p=.08) and minimum temperature (ρ=.26, p=.08). When the causes of death are divided into cancer mortality, mortality from cardiovascular disease, and mortality from other natural and non-natural diseases, one finds that the already low correlation results mostly from mortality related to cardiovascular disease. Figure 4.1 shows the scatterplots of the peak-to-trough differences in mean age at death by month of birth and the four variables.

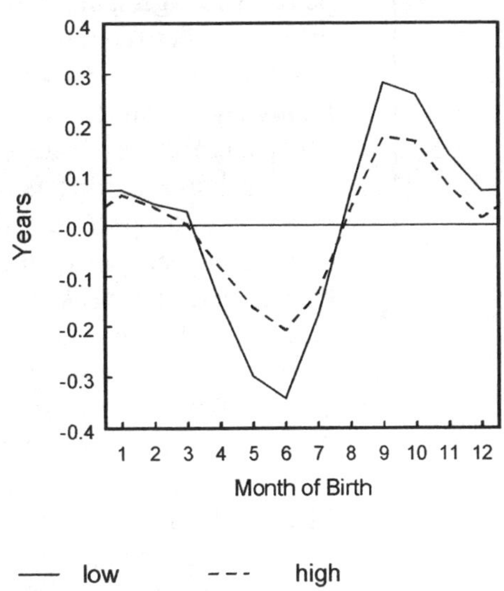

Figure 4.2. Education and the differences in mean age at death by month of birth, US death records 1989 to 1997.

Education

Comparing the two educational groups one finds that the mean age at death of those with a basic education is 78.35 years, while it is 75.43 years for people with a high level of education. This is contrary to the evidence from a large body of literature stemming from research over the last three decades which proves that the lower educated have shorter life spans than the higher educated (Kitagawa & Hauser 1973, Valkonen 1989, Elo & Preston 1996). In the cross-sectional death data, the higher mean age at death of the lower educated is caused by a change in the educational distribution over time. The oldest decedents are more likely to have received basic education than those who died in their 60s or 70s. Sex differences in education are another possible explanation: women have low education but high life expectancy.

Figure 4.2 shows that the age-standardised differences in life span by month of birth vary significantly ($p<.0001$) according to education levels. The difference between the spring trough and the winter peak is 0.62 years (p-value $< .0001$) for those with a low education and 0.38 years for the highly educated (p-value $<.0001$).

Race

The month-of-birth pattern of the 1.7 million blacks with a mean age at death of 73.2 years differs significantly ($p<.0001$) from that of the 13.9 million whites, who died at an average age of 77.03 years (Fig. 4.3). The pattern for whites reveals the familiar features: a sinusoidal curve with a June trough and a September peak. The age-standardized difference between the two months is 0.49 years and is highly significant ($p<.0001$). The pattern for the black decedents, however, differs not only with respect to the number of years between the trough and the peak (0.57 years, $p<.001$) but also with respect to the shape of the curve. The mean age at death is highest for those born between January and March. As in the case of whites, the mean age at death is lowest for decedents born in July.

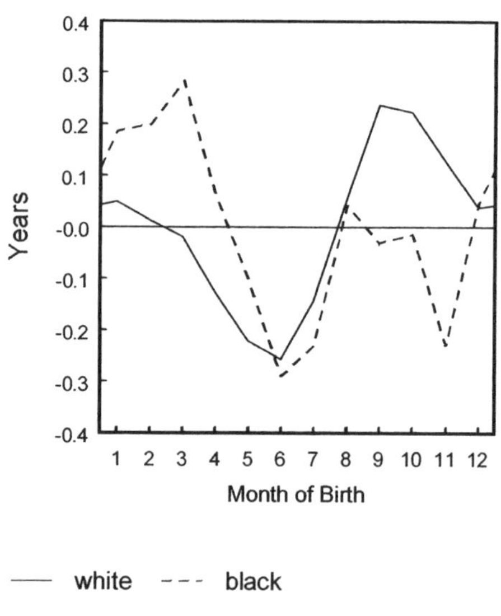

Figure 4.3. Race and the differences in mean age at death by month of birth, US death records 1989 to 1997.

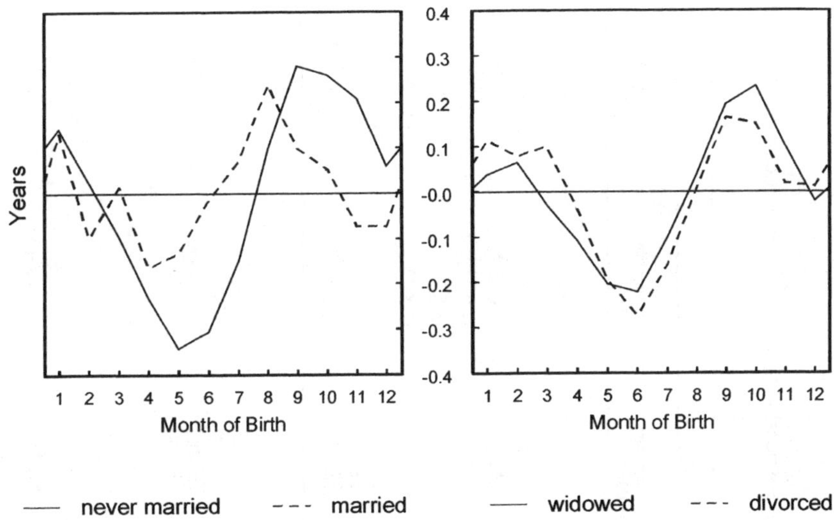

Figure 4.4. Deviation of mean age at death of decedents born in a specific month from average mean age at death by marital status: US death records 1989 to 1997, ages 50–100.

Marital Status

There exists a highly significant difference ($p<.0001$) in the month-of-birth pattern by marital status. As expected, the age-standardized difference between the peak and the trough is largest for the never-married (0.62 of a year, $p<.0001$), and smallest for the married (0.40 of a year, $p<.0001$); the widowed (0.45, $p<.0001$) and the divorced (0.44, $p<.0001$) are intermediate (Fig. 4.4).

Region of Birth

The age-standardised peak-to-trough difference in the month-of-birth pattern increases from the North to the South. The difference is smallest in New England, with 0.31 years and largest in the East South Central Region, with 0.86 years (Table 4.2). The differences in the West are intermediate. The pattern is similar in all regions with a trough in June/July and a peak in September/October. Most regions reveal a secondary peak in the first three months of the year.

Tabelle 4.2. Peak-to-trough difference in mean age at death of decedents born in a specific month by race, education and birth region.

	Age-standardised difference in years					
	US total		Race		Education	
	Peak/ trough	Max difference	Black	White	Low	High
Region of birth						
New England	Sep/Jun	0.31	1.47	0.31	0.45	0.35
Middle Atlantic	Sep/Jun	0.36	0.53	0.37	0.47	0.36
East North Central	Sep/Jun	0.46	0.81	0.45	0.65**	0.34**
West North Central	Oct/Jul	0.44	0.84	0.44	0.67**	0.37**
South Atlantic	Oct/Jun	0.46	0.65++	0.63++	0.53*	0.57*
East South Central	Oct/Jun	0.86	0.68++	1.02++	0.89*	0.81*
West South Central	Oct/Jun	0.69	0.97++	0.77++	0.89**	0.58**
Mountain	Oct/May	0.46	2.60	0.55	0.61	0.53
Pacific	Sep/Jun	0.42	1.68	0.42	0.69	0.33

All peak-to-trough differences are significant at $p<.0001$
++ indicates significant racial differentials at $p<.0001$
** indicates significant educational differentials at $p<.0001$
* indicates significant educational differentials at $p=.05$

Multivariate Results

The differences in the month-of-birth pattern depend on education, race, marital status, and region of birth. In the multivariate analysis of ages at death, the main effects of sex, month of birth, education, race, region of birth, marital status and all the two-way interactions of the variables are highly significant at p<.0001. Overall, the model explains 25 per cent of the variation in ages at death. About 86 percent of the model explanation is due to the effect of marital status and 0.4 percent to the effect of month of birth. The large majority (70%) of this 0.4 percent result from the interactions of month of birth with region of birth and race (Table 4.3). This implies that the regional differences in the month-of-birth pattern are neither due to differences in education (Fig. 4.5) nor to differences in race (Fig. 4.6) but they exist independently of them.

The highly significant racial differences (p<.0001) in the month-of-birth pattern stem solely from the South. In the North the pattern does not differ by race. In the South the pattern for whites follows a sinusoidal curve with a peak in October and a trough in June/July. The curve for blacks has its major peak in the first three months of the year and its trough in June/July. Contrary to the case of whites, blacks born in the autumn do not experience a major mortality advantage later in life.

Table 4.3. Contribution of the interactions between month of birth and sex, education, race, marital status, and region of birth to the overall model explanation of the month-of-birth effect of 0.4 per cent.

Month of Birth (MoB) effect	Per cent	p-value
MoB	3.3	<.0001
MoB*sex	1.0	<.0001
MoB*education	8.6	<.0001
MoB*race	33.2	<.0001
MoB*marital status	17.4	<.0001
MoB*region of birth	36.5	<.0001
Total contribution of the month of birth effect (0.4%) to the overall model explanation of 25%	100.0	<.0001

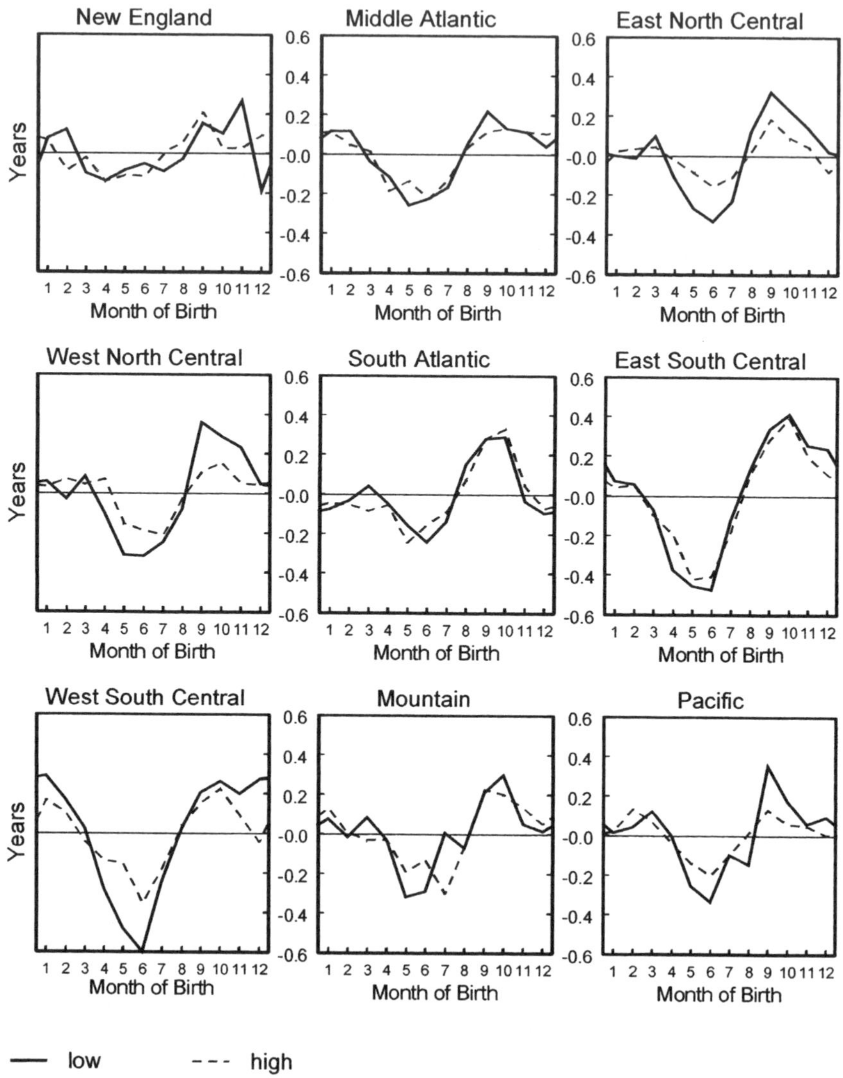

Figure 4.5. Deviation in mean age at death of decedents born in a specific month from average mean age at death by region of birth and education, US death records 1989 to 1997, ages 50-100.

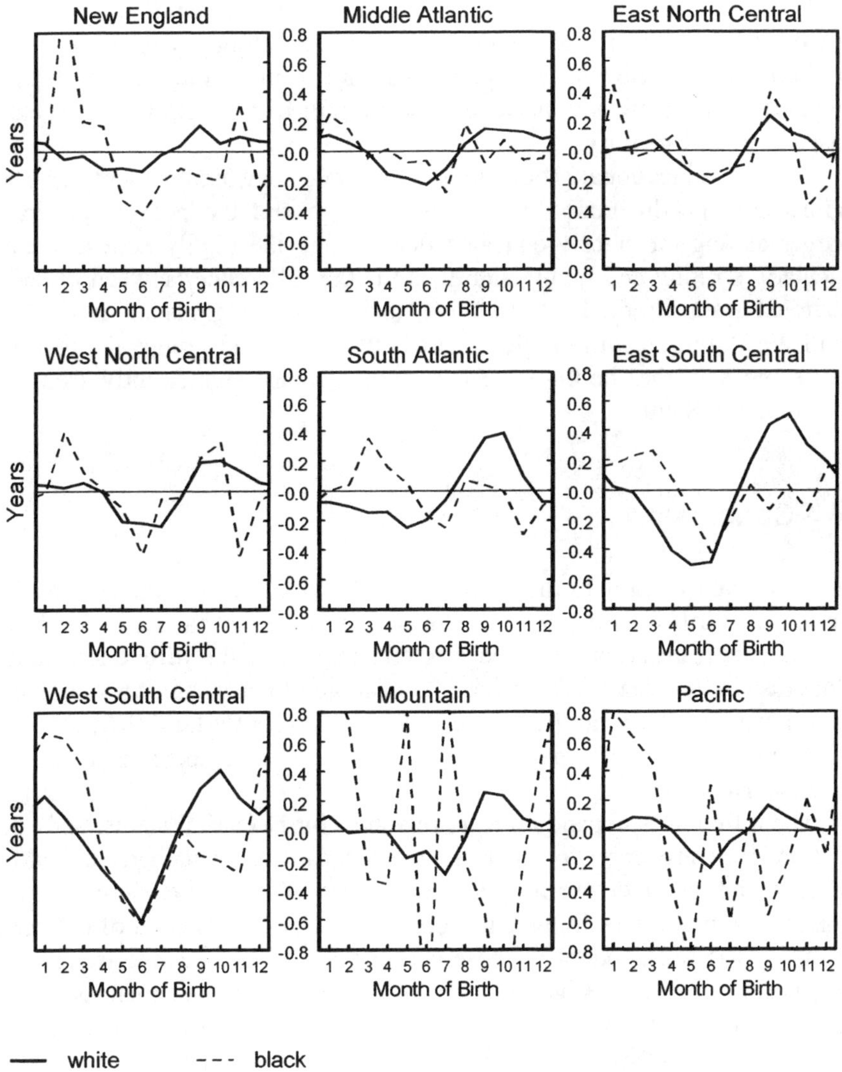

Figure 4.6. Deviation in mean age at death of decedents born in a specific month from average mean age at death by region of birth and race, US death records 1989 to 1997, ages 50-100.

Contrary to the case of whites, blacks born in the autumn do not experience a major mortality advantage later in life. Small numbers cause the erratic month-of-birth patterns for blacks born in the Mountain and Pacific region, since hardly any blacks were born in these two regions in the early part of the 20th century.

The two educational groups follow the same sinusoidal pattern in all regions and the difference between the trough and the peak is generally larger among the poorly educated than among the highly educated. Particularly striking and highly significant is the difference between the educational groups in the East and West North Central regions and in the East and West South Central regions. For both educational groups the regional differences in the month-of-birth pattern increase significantly from the North to the South.

4.4 Conclusion

A significant maximum difference of 0.44 years in mean age at death by month of birth is found for the United States on the basis of 15 million death certificates. This difference is comparable to the differences found for Austria (0.6 years), Denmark (0.3 years) and Australia (0.6 years), but it is much smaller than the 2.6 years that was recently published for Kiev (Vaiserman 2002). The pattern of the differences resembles those of Austria and Denmark.

The differences depend on race, region of birth, marital status, and education. They are larger for the less educated, for the never-married, and for blacks, and they increase from the North to the South. This increase is not caused by differences in the educational or racial composition of the decedents – it is independent of these characteristics. The peak-to-trough difference in life span is slightly larger for African-Americans than for whites and the pattern follows a different curve. This difference in the pattern, however, is only visible in the South.

The month-of-birth effect is largest among the never-married, which supports the view that the month of birth is related to the frailty of the population. There has been ample research to show that there is health selection into marriage (Lillard & Panis 1996) and that the never-married experience higher mortality (Goldman & Hu 1993).

There is a negative correlation between the maximum differences in the month-of-birth pattern and the latitude of state of birth. This allows us to rule out the hypothesis that the differences are related to the seasonality in the duration of sunshine through its effect on neuroendocrine functions. To accept this hypothesis, a positive correlation would have been necessary. The hours of duration of sunshine are more equally distributed over the year at lower latitudes than at higher latitudes. One would therefore expect that the peak-to-trough differences in mean age at death are smaller at lower latitudes than at higher.

The trough in life expectancy for those born in June and July rather than in April and May allows us to reject the hypothesis that high temperature at the time of conception is the responsible factor. Those born in June and July were conceived in September and October, which lies between one and three months after the period of maximum temperatures (July and August).

The rejection of these two hypotheses is consistent with the outcome of a review of more than 250 studies about schizophrenia and bipolar disorder (Torrey et al. 1997, Davies et al. 2003). The studies that tested the existence of a correlation between latitude and the month-of-birth effect came to mixed conclusions: some suggested the existence of a latitude gradient (Dalen 1975, Torrey 1977) while others did not (Torrey et al. 1991). McNeil et al. (1975) were unable to find a correlation between summer temperatures at the time of conception and winter births of individuals with schizophrenia. Hare and Moran (1981) and Watson et al. (1984) did not find a correlation between summer temperatures and winter births. This finding does not exclude the possibility that temperature or duration of sunshine in the first period of life has an effect on one's susceptibility to very specific diseases later in life. However, these diseases are of minor importance with respect to life expectancy. In order to affect life expectancy, major groups of causes of death such as malignant neoplasms and heart disease would need to be correlated with latitude or duration of sunshine, which is clearly not the case in the US.

The regional differences in the month-of-birth pattern are consistent with the hypothesis that nutrition and infectious disease *in-utero* or in the first year of life are the causal mechanisms. The amplitude of the month-of-birth pattern is largest in the South, which is the US region where in the first half of the 20th century diet, particularly in rural areas, was deficient throughout the year and where malnutrition was common in winter and early spring. It is also the part of the US with the highest incidence of infectious diseases, particularly of the gastro-intestinal tract, due to the hot and humid climate.

The multivariate model of age at death shows that educational differentials in the month-of-birth pattern are small compared to racial or regional differences. This finding is further support for the hypothesis that the factors that are mainly responsible for infant mortality in general are also the factors that cause the differences in life span by month of birth. The second chapter of this monograph shows the high level of positive correlation between the month-of-birth pattern in infant mortality and in life expectancy. From Preston and Haines we know that the three single most important factors explaining most of the variance in US child mortality at the beginning of he 20th century were race, size of the birthplace, and region (Preston & Haines 1991). The contemporary view at the beginning of the 20th century was that the mother's health status and whether she breastfed her baby were two factors – in addition to housing, sanitation, and general poverty – that were closely linked to the survival of infants. In other words, infant mortality was largely determined by the nutrition and health of the mother and by breastfeeding, which helped prevent infectious disease in the first months of life.

In the present analysis, the size of the birthplace is missing, which may explain the puzzling finding that the month-of-birth pattern among African-Americans is not much larger than among the white population. African-Americans, particularly in the South, lived mainly in rural areas, where life circumstances were healthier than in urban areas. Among the childhood conditions that predicted survival to advanced ages among African-Americans, the factors that were found to be most predictive were having a farm background, literate parents, and living in a two-parent household (Preston, Hill & Drevenstedt 1998). In other words, correcting for the size of the place of birth may significantly increase the month-of birth effect for African-Americans.

The difference between an urban and a rural environment may also account for the finding that the pattern differs between blacks and whites in the US South. One explanation is that the hard farm-work in the second part of the year prevents African-American infants born in the autumn from experiencing mortality advantages similar to those of the white population. Peak growth *in-utero* is during the third trimester. For infants born in the autumn, this peak growth coincides with a period of heavy work and therefore of high energy expenditures for their pregnant mothers. Evidence from Gambia shows that the high level of energy expenditure and low level of energy intake on the part of pregnant mothers during the last trimester result in a peak-to-trough difference in birthweight of approximately 200-300g (Moore 1998) and in a significant difference in survival to age 45 by month of birth (Moore et al. 1997). A heavy workload on the fields would also prevent mothers from breastfeeding their infants and would expose the infants to a higher risk of infectious disease. In Chapter 7 we further explore the explanation that the poverty of African-Americans in the South at the beginning of the 20th century together with their specific seasonal agricultural workload prevented the autumn-born from experiencing a mortality advantage similar to that of the white population.

Recent research about human skin coloration and vitamin D production points to another possible explanation why the month-of-birth patterns of whites and blacks differ. A recent study (Jablonski & Chaplin 2000) shows that the skin color graduation of indigenous peoples is highly correlated with UV radiation levels. The skin coloration is the compromise between the conflicting requirements of protection of the skin from UV-induced injuries and vitamin D synthesis, which depends on sunlight. Indigenous tropical peoples have more highly melanized skins, which provide better protection against the deleterious effects of UV radiation, such as sunburn and skin cancer. The more lightly pigmented skins of peoples inhabiting higher latitudes are thought as adaptation to lower UV radiation and the importance of maintaining UV induced biosynthesis of vitamin D_3 in the skin. Increasing the melanin in the human skin increases the length of exposure to sunlight that is needed for the production of the previtamin D_3. If the duration of UV exposure of deeply melanized skin is not sufficient then individuals are at much higher risk of vitamin D_3 deficiency and its manifestations in the form of certain diseases. Research findings from migrants from Ethiopia to Israel (Fogelman et al. 1995) showed a high prevalence of vitamin D deficiency among female migrants and highlighted the importance of limited exposure to UV radiation in the development of rickets among migrants from the Indian subcontinent to urban centers in the UK (Henderson et al. 1987).

Vitamin D deficiency in the earliest periods of life has long-term effects. A study based on sera taken during the third trimester of pregnant women found lower levels of vitamin D among black women and an increased risk of schizophrenia among the children of these women (McGrath et al. 2003). An ecological analysis of long-term trends in schizophrenia in the Netherlands and in Australia showed that individuals born during periods of decreased perinatal sunshine are at a higher risk of developing schizophrenia (McGrath et al. 2002). Vitamin D deficiency has been suggested to be related with a variety of diseases, such as schizophrenia, multiple sclerosis, Parkinson's disease, Alzheimer's disease, diabetes I, high blood pressure, heart disease and breast cancer, bowel cancer and ovary cancer.

According to above research African-Americans born in the US are at a higher risk of vitamin D deficiency. The risk may be increased for the fall-born either due to vitamin D deficiency of the mother during pregnancy and lactation or due to a lack of exposure to sunshine of the infant during its first months of life.

5 Month of Birth and Causes of Death in the United States

Since US death records contain more then 15 million decedents, it is possible to verify the month-of-birth pattern for the same causes of death that were analysed in the second chapter on the basis of the Austrian data. In general, the month-of-birth patterns for Austria are confirmed. However, those causes of death that were of borderline significance or insignificant in the Austrian study are now significant in the US data set (e.g. lung cancer, diabetes mellitus).

The following part of this chapter presents the results for different causes of death and relates the findings to existing research. Some literature exists about the relationship between month of birth and the incidence of malignant neoplasms. Extensive research has been done on nervous and mental disorders. Hardly any studies exist, however, about the relationship between month of birth and circulatory disease.

Although the dataset consists of more than 15 million decedents, the month-of-birth patterns for some of the causes of death tend to be erratic, particularly for African-Americans. To strategies are followed in the analysis of the data. Regression models similar to Equ. 3.1 are estimated for each cause of death in order to test whether the month-of-birth patterns differ significantly by sex, race, education and region. The results are reported in Table 5.1 at the end of this chapter.

Cosinor analysis is applied to test whether the month-of-birth pattern follows a sinusoidal fluctuation. In particular, for each cause of death three models are estimated on the basis of the individual death records: the first model tests whether an unimodal sinusoidal function fits the data and defines the period 2π in terms of 12 months (i.e. p=12); the second model tests for a bimodal sinusoidal function and assumes that the period is defined over p=6 months.

$$y = \alpha_0 + \alpha_1 \cos(t) + \alpha_2 \sin(t) + u$$
$$t = month\ of\ birth / p * 2\pi \qquad [5.1]$$

The third model tests whether the sum of a unimodal and a bimodal sinusoidal function fits the data best. Equ. 5.1a allows for modeling of seasonal fluctuations that depart from pure sinusoidal functions. The final model is selected on the basis of the Akaike information criteria max (2LL-2p), where LL denotes the log likelihood and p the number of parameters in the model. Parameter estimates are reported in Table 5.2 at the end of this chapter. The mean ages at death by month of birth together with the estimated sinusoidal functions and the 95% confidence bands are displayed in figures 5.1 to 5.8.

$$
\begin{aligned}
y &= \alpha_0 + \alpha_1 \cos(t) + \alpha_2 \sin(t) + \alpha_3 \cos(t_1) + \alpha_4 \sin(t_1) + u \\
t &= month\ of\ birth / p * 2\pi \\
t_1 &= month\ of\ birth / \frac{p}{2} * 2\pi
\end{aligned}
\qquad [5.1a]
$$

5.1 Malignant Neoplasms

There exist highly significant differences (p<=.001) in mean age at death by month of birth for breast cancer, prostate cancer, lung cancer, colorectal cancer and the residual group of other cancers. Differences are significant at p=.05 for stomach cancer, pancreatic cancer and liver cancer, and not significant (p>.05) for uterus cancer. None of the cancers reveal differences by sex or educational status (Table 5.1, Figs. 5.1 and 5.2).

Lung cancer is the only malignant neoplasm with significant differences by race. Similarly to total mortality, mean age at death for blacks peaks among the March-born and does not reveal an autumn peak as is the case for whites. In contrast to total mortality, there is no North-South trend in the magnitude of the difference. Differences are smallest in the Middle Atlantic region and seem to be largest in the Mountain region. It is not only the magnitude that changes between the regions but also the shape of the curve, with the East South Central region revealing a particularly distinct pattern. In addition to breast and lung cancer, the month-of-birth pattern of stomach, colorectal, and the residual group of other cancers also differ significantly by region of birth.

These results are consistent with those of Jansson and Malahy (1981), which are based on the Third National Cancer Survey of the United States, consisting of about 180,000 patients. They find that patients born between January and April are diagnosed with cancer 1–1.5 years later and live one year longer than patients born from June to August. They do not find differences in the season-of-birth distributions of cancer patients between males and females. They find differences in the month-of-birth distribution for different sites of the lesions: colon, rectum and breast cancer patients have an almost identical distribution, while stomach and lung cancer patients have significantly different season-of-birth distributions. The authors attribute this to differences in the geographic distributions and to social differences.

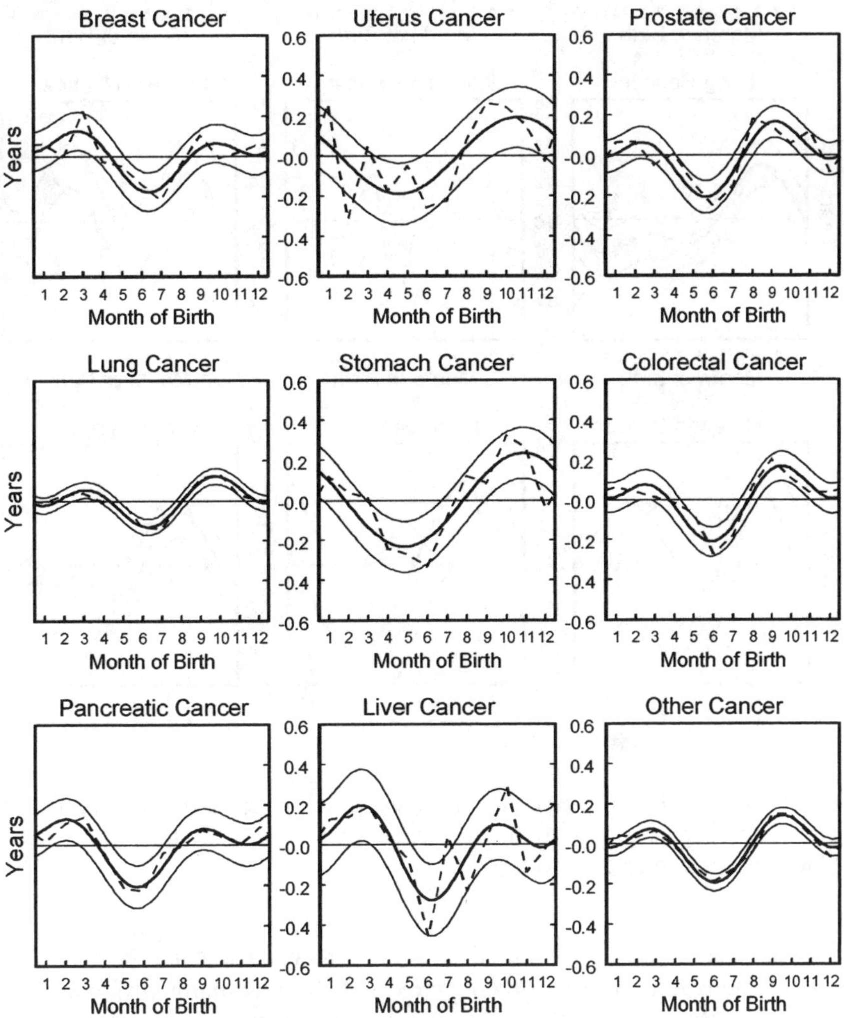

Figure 5.1. Malignant neoplasms: deviation in mean age at death of decedents born in a specific month from the average mean age at death, United States whites, death records 1989 to 1997.

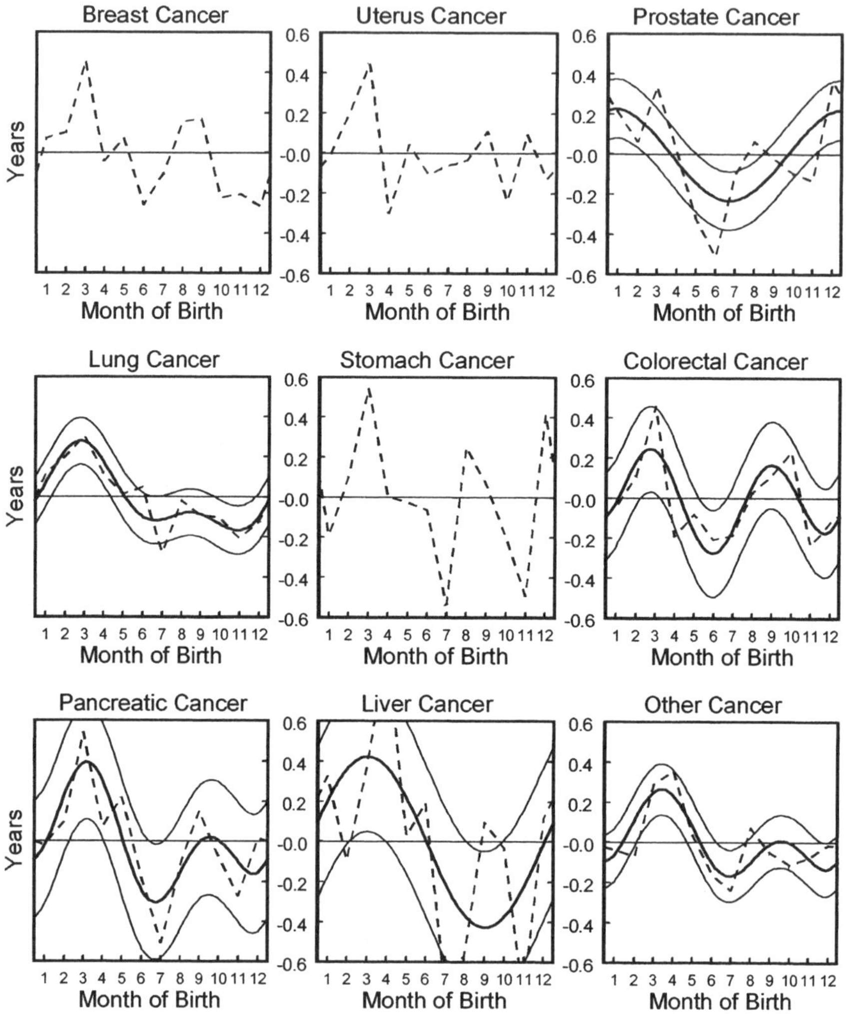

Figure 5.2. Malignant neoplasms: deviation in mean age at death of decedents born in a specific month from the average mean age at death, United States blacks, death records 1989 to 1997.

When the authors compared the maximum difference between the monthly temperature at a given place and the maximum difference between monthly frequencies of birth among cancer patients they find a positive correlation of r=0.89. This is different from the present finding that only a small correlation exists between the maximum mean age at death by month of birth and temperature variables. Jansons and Malahy's approach, however, confines monthly differences in the seasonal distribution of the number of births with the month-of-birth effect on life span. Thus, the high correlation is most likely the result of the correlation between the seasonal distribution of births and temperature, which has been demonstrated by Lam and Miron (1996), among others. The authors conclude that the relationship between month of birth and cancer probably reflects general differences in the health status of persons born at different times of the year rather than being related to the aetiology of cancer.

A Swedish study compares the birth distribution of 115,670 women with breast cancer to the total number of live births by month of birth from 1858 to 1968 (Yuen et al. 1994). The month of birth was a significant risk factor for breast cancer, but not for other types of cancer (440,948 women). For the older birth cohorts (1880–1920) they find an increased risk for women born in May and June. The pattern for women born after 1950, however, was completely different, with a maximum in autumn. Age and cohort are confounded in their analysis, since these women had a maximum age of 39 and thus represent pre-menopausal breast cancer, whose etiology may differ from breast cancer in post-menopausal women. The authors discuss theories of seasonal fluctuations in pregnancy estrogen, other pregnancy hormones, and post-natal exposures such as diet. They also mention seasonal differences in birth weight as a risk for breast cancer.

Differences in the results regarding the incidence of pre- and post-menopausal breast cancer are also reported in an earlier study by Nakao (1988). A Japanese study (Hu et al. 1996) did not find any differences in the breast cancer risk according to the month of birth for 81,162 women who died of breast cancer.

A series of studies about testicular cancer was conducted in the late 80s. Results, however, are not conclusive. Two studies found significant bimonthly cycles (Prener & Carstensen 1990, Kinlen & Willos 1987), and one found a significant 4-month cycle (Knox & Cummins 1985). Bernstein et al. (1986) observed a significant annual cycle with a distinct peak in August, confined to teratomas.

Three recent studies describe month-of-birth patterns in childhood leukemia (Sorensen et al. 2001, Higgins et al. 2001, Feltbower et al. 2001) with an increased risk for children born in winter and early spring. Infectious disease is discussed as the causal factor.

5.2 Diseases of the Circulatory System

The differences in life span by month of birth result largely from circulatory diseases (Table 5.1). Acute myocardial infarctions, ischemic heart disease, cerebrovascular disease, the residual group of other heart diseases, as well as the causes of death related to circulatory diseases such as accidental falls, diabetes mellitus (ICD9 250), and pneumonia and influenza show a distinct month-of-birth pattern (Figs. 5.3 and 5.4). People born in the spring die earlier than those born in the autumn. There are no differences by sex, but there are highly significant differences by race. Among blacks the peak in mean age at death is shifted to the first three months of the year. In addition to the differences by race, there exist highly significant differences by marital status, with the never-married born in May and June being particularly disadvantaged. In addition, highly significant regional differences do exist and the month-of-birth effect is strongest in the East South Central and the West South Central regions.

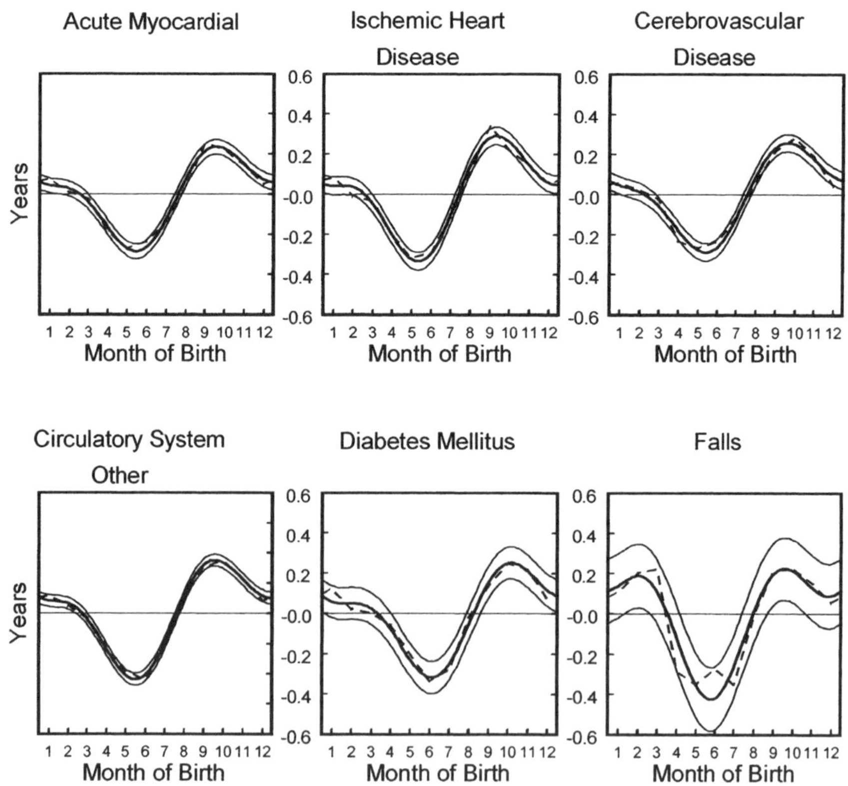

Figure 5.3. Heart disease: deviations in mean age at death for decedents born in a specific month from the average mean age at death, United States whites, death records 1989 to 1997.

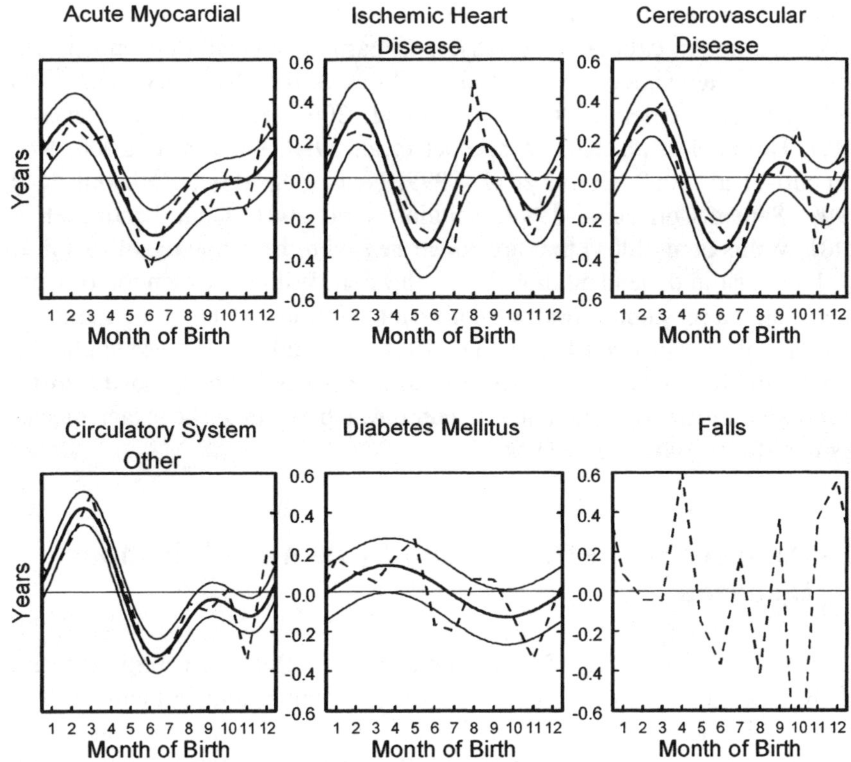

Figure 5.4. Heart disease: deviations in mean age at death for people born in a specific month from the average mean age at death, United States blacks, death records 1989 to 1997.

Most research that explores the effect of early-life circumstances on late-life health and mortality focuses on circulatory diseases (for reviews see Elo & Preston 1992, Kuh & Davey Smith 1993; Barker 1994, Barker 1995, Davey Smith et al. 1997, Davey Smith et al. 2000). Surprisingly, only two studies exist on how these diseases are related to the month of birth. They both explore the relationship between month of birth and death by cerebral haemorrhage. Caroll & Haddon (1964) included deaths under age 34 and did not find any relationship. Nonaka & Imaizumi (2000) studied subarachnaoid haemorrhage (ICD9 code 430) in Japan by comparing the month-of-birth distribution on death records from the years 1986–1994 with the monthly distribution of births for the years 1900–1959. They found an excess risk of 8 to 23 per cent for people born from June to September. More attention has been paid to diseases that are considered to be a risk factor for circulatory diseases, in particular insulin-dependent diabetes. There are sixteen recent studies (Block et al. 1994, Chen et al. 1998, Dahlquist & Kaellen 1996, Fichera et al. 2001, Helgasson & Jonasson 1981, Hummel et al. 2001, Jongbloet et al. 1998, Kida et al. 2000, Kordonouri et al. 2002, Laron et al. 1999, Neu et al. 2000, Rothwell et al. 1999, Samuelsson et al. 1999, Songini et al. 2001, Ursic-Bratina et al. 2001, Willis et al. 2001) that are concerned with the influence of month of birth on insulin dependent diabetes in children. With the exception of Chen et al., all these studies find a significant relationship between month of birth and the incidence of type I Diabetes. The authors of these studies attribute this relationship primarily to intrauterine infections, mostly during the yearly influenza epidemics in autumn and winter. The exact mechanism is still unknown, however.

5.3 Diseases of the Respiratory System and Infectious Diseases

There exists a highly significant month-of-birth effect for chronic diseases of the respiratory system as well as for pneumonia and influenza (Table 5.1, Fig. 5.5 and Fig. 5.6). The effect differs by region and race and in the case of chronic respiratory diseases also by marital status. Considerable evidence exists that lower respiratory tract infections early in life lead to chronic obstructive lung disease late in life (for a review see Elo & Preston 1992). In young children, viruses are the principal etiological agents that cause lower respiratory tract infections. Elo and Preston point out that, since the lungs of children undergo developmental changes, it is plausible that clinically severe infections could have lasting effects. A large amount of literature exists which shows a significant relationship between the month of birth and atopic disease in childhood. One frequently explored disease is wheezing and asthma.

A study conducted in Queensland, Australia, shows that school children who experience frequent wheezing (one or more episodes per month) are 60% more likely to be born in spring or summer (Duffy 1991). In Hong Kong the prevalence of wheezing is highest for girls born in January/February and lowest for those born in July/August (Lau & Karlberg 1998). In Singapore adolescents born in two main seasons, namely March–May and September–November, show a higher prevalence of asthma and asthma-like diseases (Chew et al. 1998).

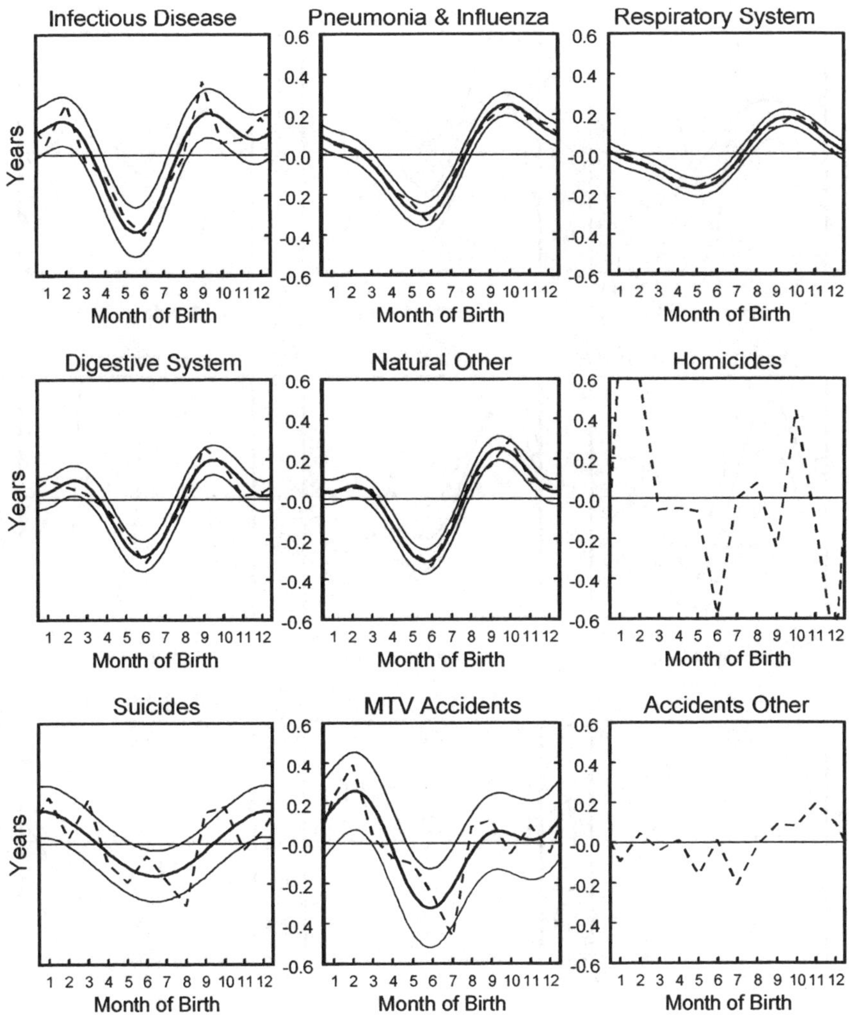

Figure 5.5. Natural and non-natural causes of death: deviation in mean age at death for decedents born in a specific month from the average mean age at death, United States whites, death records 1989 to 1997.

A series of studies explores IgE antibodies. A Swedish study (Nielsson et al. 1997) found that children aged 0–15 born during the tree pollen season (spring) were less likely to develop allergic rhinoconjunctivits, IgE antibodies to pollen, or a positive screening test for IgE antibodies than

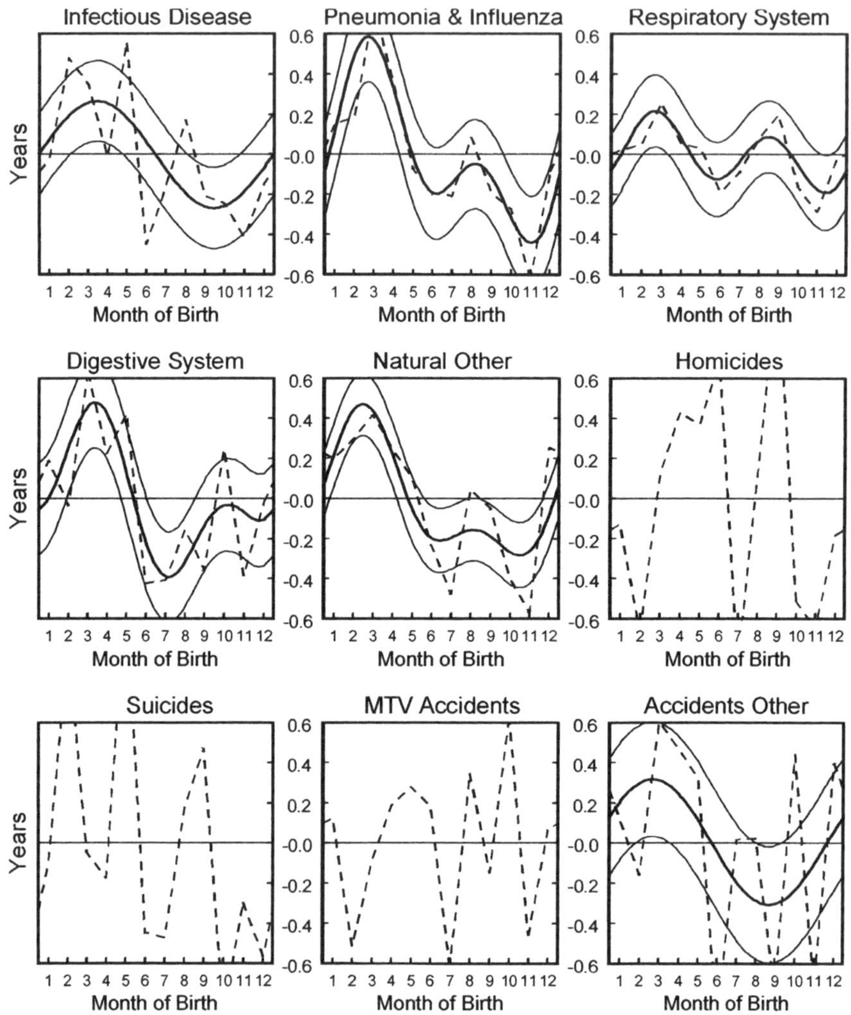

Figure 5.6. Natural and non-natural causes of death: deviation in mean age at death of decedents born in a specific month from the average mean age at death, United States blacks, death records 1989 to 1997.

children born throughout the rest of the year. The prevalence of IgE antibodies to food and animal dander was higher in children born in autumn and winter (September to February). An earlier Swedish study (Eriksson & Holmen 1996) on adults (ages 14+) came to slightly different results, however. Those born between February and May had a higher prevalence of isolated pollen allergy. A Dutch study (Aalberse et al. 1992) suggests that for the three seasonal allergens birch pollen, grass pollen, and house dust mite an increasingly elevated risk exists for those born up to three months before the main season of the allergen. For cat and dog allergies, those born between November and January had an increased risk as compared to the May-born. The authors conclude that a sensitive period exists in the first months, during which allergy exposure is more likely to prime for an allergy later in life. A French study (de Montis 1998) confirms the results concerning house dust mite allergy and finds a peak for people born in October and a trough for those born in April. Further evidence for there being a significant relationship between allergies to house dust mites and pollen and month of birth comes from a Turkish study (Erel et al. 1998). A German study, however, does not find a correlation between allergic sensitisation and manifest atopic disease and month of birth (Schaefer et al. 1993).

While the studies about the prevalence of wheezing and asthma in childhood by month of birth suggest that the spring-born have a higher risk of atopic manifestation, a study of lower respiratory tract infections in childhood comes to the opposite conclusion. The study is based on British school children and middle-aged adults, and it finds that admission to a hospital for bronchiolitis in the first year of life was three times more common among children born from September to November than among those born from March to May. Admissions for asthma were significantly more common among children and young adults born in the autumn. Among adults however, those born in autumn had a higher forced ventilatory capacity and there was little variation in hospital admissions for chronic bronchitis/emphysema and pneumonia, although there was a tendency for those born in the spring to have higher admission rates (Strachan et al. 1994). The authors conclude that there is no evidence in their study for a causal link between chest illness in infancy and the later development of chronic bronchitis and emphysema.

Significant differences in the month-of-birth effect also exist for mortality from infectious diseases in less developed countries. A recent study of the relationship between month of birth and mortality in Gambia found that those born shortly after the hunger season had a higher risk of dying before their 45th birthday (Moore 1997). All the deaths were caused either by pregnancy-related deaths or by infectious disease. The authors conclude that malnutrition together with a high level of energy expenditure on the part of mothers in their last trimester of pregnancy affected the immune systems of their babies.

Two studies explore the relationship between month of birth and Crohn's disease. They are based on a recent finding that infections early in life or in-utero – measles in particular – seem to be important risk factors for the development of Crohn's disease later in life (Ekbom et al. 1996). However, other studies failed to find a relationship (Nielsen et al. 1998). Since the incidence of measles and other possible infectious agents is seasonal, the seasonal variation might be reflected in the pattern by month of birth in people with Crohn's disease. Results from the two studies exploring the month-of-birth effect are not consistent. In Britain people born in the first half of the year seem to have a slightly increased risk of Crohn's disease (Haslam et al. 2000), whereas in Denmark the peak in prevalence is observed for people born in August (Toft-Sorensen et al. 2001).

A recent study (Ivarsson et al. 2003) explores the month-of-birth pattern of coelic disease in Swedish children below age 15. For children under age two there is an excess of children born during the summer. For children aged two to 15 no pattern exists. The authors suggest that infections may cause the month-of-birth pattern.

5.4 Non-Natural Causes of Death

There exist significant differences in mean age at death by month of birth for suicides, car accidents (MTV), and for the residual group of all other non-natural causes of death (Table 5.2, Fig. 5.5 and Fig. 5.6). Those born in the spring experience a higher risk than those born in the autumn. Con– sidering the large number of deaths that fall into each of these groups, the month-of-birth effect is only of borderline significance.

The evidence concerning a month-of-birth effect for suicide is mixed. A recent study on Alaskan natives showed more suicides among summer births (Kettl et al. 1997). Likewise, Pokorny (1960) found an overrepresentation of suicide cases among people born in July; Salib (2002) of those born in May. Three other studies could not find a relationship (Lester 1987, Lester et al. 1970, Sanborn & Sanborn 1974). Although the relationship between month of birth and the risk of suicide may be surprising at first glance, plausible explanations exist. Suicide at old age is often related to chronic diseases, which according to our results are significantly influenced by the month of birth. Chotai et al. (1999) report differences in suicide methods according to month of birth. Those who preferred hanging rather than poisoning or carbonmonoxide poisoning from a running motor were significantly more likely to be born in February–April than October–January, particularly among males aged 45 or younger. Salib (2002) reports similar findings. The suicide attempts may thus be independent of the month of birth. Those born in late winter/early spring, however, may use suicide methods that lead to death more often than those born in the autumn. This question should lead to research into the effect of month of birth on personality, which, due to the extensive nature thereof, is beyond the scope of this chapter.

Differences in mean age at death by month of birth for car accidents may be explained by the month-of-birth pattern for alcoholism that has been demonstrated repeatedly (Goldberg & Newlin 2000, Modestin et al. 1995, Levine & Wojcik 1999, London 1998, Kunugi et al. 1998, Lang 1931) although two other studies did not find any pattern (Dalen 1975, Watson et al. 1984).

5.5 Diseases of the Nervous System and Psychiatric Disorders

Most of the research on the relationship between month of birth and the incidence of diseases has been conducted for mental disorders, in particular schizophrenia and other bipolar disorders (for reviews see Davies et al. 2003, McGrath & Welham 1999, Torrey et al. 1997, Fossey & Shapiro 1992, Bradbury & Miller 1985). Torrey et al. review more than 250 studies, covering 29 countries from the northern hemisphere and 5 from the southern hemisphere, and find a remarkable consistency in the results. There is a 5 to 8 per cent winter/spring excess of births for both schizophrenia and mania/bipolar disorders. The season-of-birth effect has also been studied for autistic disorder (Barak et al. 1995, Bolton et al. 1992, Ticher et al. 1996), Alzheimer patients (Philpot et al. 1989, Dysken et al. 1991, Henderon et al. 1991, Vitiello et al. 1991, Vezina et al. 1996), and for anorexia nervosa patients (Nielsen 1992, Rezaul et al. 1996, Eagles et al. 2001, Waller et al. 2002, Watkins et al. 2002, Willoughby et al. 2002, Penas-Lledo & Waller 2002). With the exception of a few studies, they all find significant differences in the birth distribution of the patients as compared to the general population or the control group. For diseases of the nervous system significant differences by month of birth have been shown for Parkinson's disease (Miura et al. 1987, Mattock et al. 1988), multiple sclerosis (Miura et al. 1987, Rothwell et al. 1999, Sadovnick et al. 1994, Salemi et al. 2000, Wiberg et al. 1994) and epilepsy (Procopio & Marriott 1998).

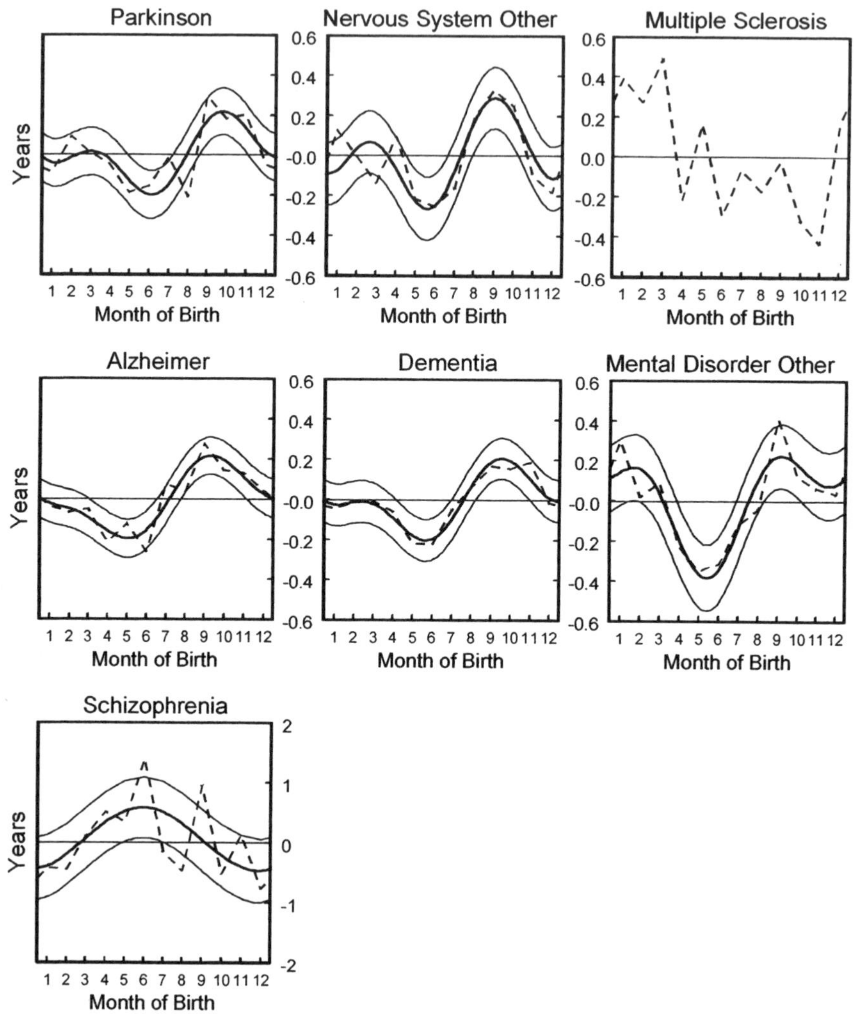

Figure 5.7. Diseases of the nervous and mental system: deviations in mean age at death for decedents born in a specific month from the average mean age at death, United States whites, death records 1989 to 1997.

The present results confirm the findings of all these studies (Table 5.3, Fig. 5.7 and Fig. 5.8) with the exception of multiple sclerosis, which may be attributable to the small number of observations in this study.

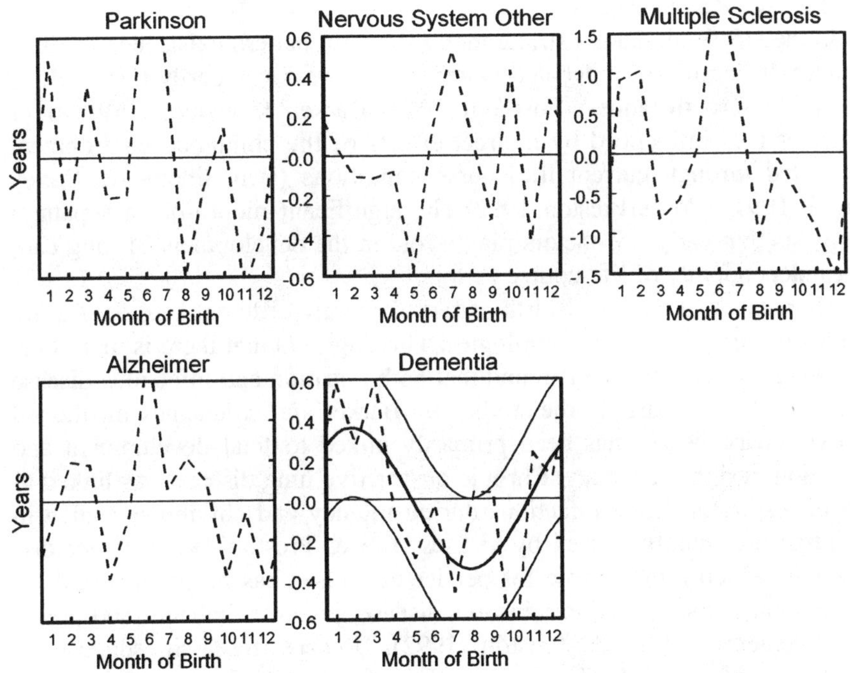

Figure 5.8. Diseases of the nervous and mental system: deviations in mean age at death for decedents born in a specific month from the average mean age at death, United States blacks, death records 1989 to 1997.

5.6 Conclusion

Heart disease, certain types of cancers, and other natural causes of death such as chronic obstructive lung disease reveal a strong month-of-birth pattern. This result is consistent with the extensive literature on the effect of early-life circumstances on late-life disease and mortality. Davey Smith and Kuh (2001) mention stroke, stomach cancer, and bronchitis as important contributors to adult mortality which declined on a cohort basis between the mid-19th century and today. Highly positive correlations between infant mortality in 1921–1925 and mortality from chronic bronchitis, ischemic heart disease, stomach cancer, and rheumatic heart disease have been demonstrated (Barker & Osmond 1986a), as well as for stroke (Barker & Osmond 1989).

In the US death data, a strong month-of-birth pattern exists also for lung cancer. It is still hotly debated whether lung cancer is partly the result of early-life circumstances (Forsdahl 1973, Barker & Osmond 1986a) or whether it is influenced by indirect effects of the childhood environment mediated through current life-course conditions (Ben Shlomo & Davey Smith 1991, Elo & Preston 1992).The significant month-of-birth pattern suggests that early life factors play a role in the development of lung cancer independently of life-course factors.

There exists a month-of-birth pattern for many different causes of death with very different disease etiologies. This suggests that there is more than one single causal mechanism and that both nutrition and infectious disease are important factors. In the studies by Barker and colleagues mentioned above, heart disease has been primarily linked to fetal development and nutrition during pregnancy. Chronic obstructive lung disease was linked to lower respiratory tract infections during infancy and childhood, which in children are mainly caused by viruses (Elo & Preston 1992). Other diseases in which a major role for persistent viruses has been suggested include multiple sclerosis, Parkinson's disease, juvenile diabetes, and presenile dementia (Elo & Preston 1992). Recent research suggests a relationship between Helicobacter pylori infections in childhood and stomach cancer (Elo & Preston 1992).

A surprising result is the similarity of the month-of-birth pattern of very different causes of death. One generally finds a trough in mean age at death for those born in spring and early summer and a mortality advantage for autumn-born whites and winter-born blacks. One possible explanation is that people born in spring and early summer are doubly disadvantaged. On the one hand, their last trimester *in-utero* falls into the season of the year where the lack of certain micronutrients such as vitamins is most severe. On the other hand, their mortality risk during their first year of life is higher than of those born during other seasons. This is an indication of the effect of infectious diseases, which largely determined infant mortality at the beginning of the 20th century.

For most causes of death, the month-of-birth pattern differs significantly between whites and blacks. In general the trough in mean age at death is similar for both groups, namely in spring/early summer. However, for African Americans the peak is shifted from autumn to winter. This means that the difference in the month-of-birth patterns for the two groups in total mortality is not caused by one or two major groups of causes of death. Rather, it is a general feature of most causes.

The month-of-birth patterns of most cancers, circulatory diseases, pneumonia and influenza, and of diseases of the respiratory system differ significantly by region of birth (not shown). For circulatory diseases the locations of the peaks and troughs remain generally the same for all regions of birth – it is primarily the amplitude that changes. For the other causes of death the peaks and troughs also vary by region of birth. This is further indication that the causal mechanisms that underlie the month-of-birth patterns are different for different diseases.

Table 5.1. Natural causes of death: significance levels of the effect of month of birth (MoB) and its interactions with sex, education, race, marital status, and region.

Causes of death	Total	Model 1	Model 2				
		MoB	Interaction MoB with				
			sex	education	race	marital status	region
Malignant neoplasms							
Breast	299,323	<.0001					.0075
Uterus	71,035	.0782					
Prostate	272,584	<.0001					
Lung	1,175,632	<.0001			<.0001	.0041	<.0001
Stomach	93,877	.0252					.0107
Colorectal	435,972	<.0001					.0078
Pancreatic	201,918	.0443					
Liver	68,710	.0183					
Other	1,328,063	<.0001		.0158	<.0001	.0008	.0005
Circulatory disease and related diseases							
Acute myocardial inf.	1,754,092	<.0001		.0144	<.0001	<.0001	<.0001
Ischemic heart disease	1,249,519	<.0001			<.0001	<.0001	.0177
Cerebrovascular disease	1,150,444	<.0001			<.0001	.0174	.0034
Other circ. system	2,971,581	<.0001			<.0001	<.0001	<.0001
Pneumonia & influenza	591,543	<.0001		.0203	<.0001		.0228
Diabetes mellitus	400,910	<.0001	.0283	<.0001			
Falls	92,389	.0016	.0555	.0307	.0184		

Table 5.1. (continued)

Other natural causes of death							
Infectious disease	242,155	<.0001			.0011		
Respiratory system	998,388	<.0001			.0039	.0027	.0167
Digestive system	523,268	<.0001			<.0001		
Other natural causes	776,124	<.0001			<.0001		.0008
Diseases of the nervous and mental system							
Parkinson	75,016	.0011				.0003	
Multiple sclerosis	12,038	n.s.					
Other nervous system	100,482	.0017					
Schizophrenia	2,459	.0382			.0062	.0101	
Dementia	112,690	.0081				.0345	
Alzheimer	139,188	<.0001		.0118			
Other mental disorders	124,167	<.0001					
Non-natural causes of death							
Homicide	22,124	n.s.					
Suicide	89,929	.0492	.0532				
Accidents motor vehicles	71,538	.0302					
Other accidents	143,821	n.s.					
Non-natural others	5,973	.0328					

Model 1 includes the main effects of month of birth (MoB), education, race, region, marital status, and sex. All main effects are highly significant.
Model 2 includes all main effects and all interaction effects significant at a level of 0.05. The main effect of month of birth is significant for all causes of death.

Table 5.2. Parameter estimates of the cosinor functions defined in Equs. 5.1 and 5.1a for major groups of causes of death; US death certificates 1987-1998, whites.

Causes of death	12 months period		6 months period		F-value	p-value
	sinus	cosinus	sinus	cosinus		
Malignant neoplasms						
Breast	0.043	0.088 **	-0.019	-0.081 **	4.32	0.0017
Uterus	-0.128 **	0.142 **			4.46	0.0116
Prostate	-0.060 **	0.089 ***	0.018	-0.106 ***	8.82	<.0001
Lung	0.020	0.061 ***	-0.039 **	-0.070 ***	16.02	<.0001
Stomach	-0.132 **	0.193 ***			9.68	<.0001
Colorectal	-0.043	0.108 ***	-0.010	-0.104 ***	10.93	<.0001
Pancreatic	0.001	0.109 ***	0.043	-0.086 **	4.53	0.0012
Liver	0.054	0.130 **	-0.008	-0.137 **	2.95	0.0189
Other	-0.026	0.087 ***	-0.022	-0.101 ***	26.01	<.0001
Circulatory disease and related diseases						
Acute myocardial	-0.125 ***	0.170 ***	0.011	-0.092 ***	91.77	<.0001
Ischemic heart disease	-0.163 ***	0.177 ***	0.034 ***	-0.119 ***	89.91	<.0001
Cerebrovascular disease	-0.144 ***	0.177 ***	0.008	-0.087 ***	76.36	<.0001
Other circulatory disease	-0.133 ***	0.198 ***	0.016	-0.108 ***	188.12	<.0001
Pneumonia & influenza	-0.123 ***	0.202 ***	-0.003	-0.082 ***	43.81	<.0001
Diabetes mellitus	-0.064 **	0.217 ***	-0.051 **	-0.095 ***	23.30	<.0001
Falls	-0.042	0.254 ***	0.028	-0.159 ***	8.95	<.0001

Table 5.2. (continued)

Other natural causes of death						
Infectious disease	-0.065	0.222 ***	0.056	-0.137 ***	12.13	<.0001
Respiratory system	-0.128 ***	0.091 ***	0.000	-0.045 ***	34.25	<.0001
Digestive system	-0.053 **	0.150 ***	-0.001	-0.129 ***	17.83	<.0001
Other natural	-0.104 ***	0.172 ***	0.008	-0.131 ***	38.61	<.0001
Diseases of the nervous and mental system						
Parkinson	-0.077 **	0.107 ***	-0.056	-0.092 **	5.02	0.0005
Multiple sclerosis						n.s.
Other nervous system	-0.116 **	0.069	0.018	-0.178 ***	5.03	0.0004
Schizophrenia	0.011	-0.790 **			3.00	0.0500
Dementia	-0.104 ***	0.103 ***	-0.015	-0.088 ***	6.64	<.0001
Alzheimer	-0.151 ***	0.091 **	0.011	-0.062 **	61.62	<.0001
Other mental disorder	-0.091 *	0.214 ***	0.079	-0.131 **	6.91	<.0001
Non-natural causes of death						
Homicides						n.s.
Suicides	0.040	0.159 ***			4.69	<.0001
Accidents motor vehicles	0.069	0.191 ***	0.054	-0.126 **	3.77	<.0001
Other accidents						n.s.
Other non-natural	-0.197	0.472 ***	-0.468 *	0.094	2.00	0.0920

Table 5.3. Parameter estimates of the cosinor functions defined in Equs. 5.1 and 5.1a for major groups of causes of death; US death certificates 1987-1998, blacks.

Causes of death	12 months period		6 months period		F-value	p-value
	sinus	cosinus	sinus	cosinus		
Malignant neoplasms						
Breast						n.s.
Uterus						n.s.
Prostate	0.086	0.214 ***			7.04	<.0001
Lung	0.180 ***	-0.001	0.028	-0.097 ***	7.32	<.0001
Stomach						n.s.
Colorectal	0.036	0.057	0.026	-0.021 ***	2.57	0.0361
Pancreatic	0.198 **	0.040	-0.072	-0.193 **	2.45	0.0438
Liver	0.425 ***	-0.014			3.69	0.0250
Other	0.129 ***	-0.016	-0.066	-0.118 ***	5.14	0.0004
Circulatory disease and related diseases						
Acute myocardial	0.157 ***	0.175 ***	0.033	-0.102 ***	10.30	<.0001
Ischemic heart disease	0.043	0.107 **	0.159 ***	-0.191 ***	7.15	<.0001
Cerebrovascular disease	0.118 ***	0.151 ***	0.051	-0.200 ***	9.96	<.0001
Other circulatory disease	0.224 ***	0.125 ***	0.028	-0.182 ***	32.93	<.0001
Pneumonia & influenza	0.122 **	-0.044			2.61	0.0736
Diabetes mellitus						n.s.
Falls	-0.042	0.254 ***	0.028	-0.159 ***	8.95	<.0001

Table 5.3. (continued)

Other natural causes of death						
Infectious disease	-0.065	0.222 ***	0.056	-0.137 ***	12.13	<.0001
Respiratory system	-0.128 ***	0.091 ***	0.000	-0.045 ***	34.25	<.0001
Digestive system	-0.053 **	0.150 ***	-0.001	-0.129 ***	17.83	<.0001
Other natural	-0.104 ***	0.172 ***	0.008	-0.131 ***	38.61	<.0001
Diseases of the nervous and mental system						
Parkinson	-0.077 **	0.107 ***	-0.056	-0.092 **	5.02	0.0005
Multiple sclerosis						n.s.
Other nervous system	-0.116 **	0.069	0.018	-0.178 ***	5.03	0.0004
Schizophrenia	0.011	-0.790 **			3.00	0.0500
Dementia	-0.104 ***	0.103 ***	-0.015	-0.088 ***	6.64	<.0001
Alzheimer	-0.151 ***	0.091 **	0.011	-0.062 **	61.62	<.0001
Other mental disorder	-0.091 *	0.214 ***	0.079	-0.131 **	6.91	<.0001
Non-natural causes of death						
Homicides						n.s.
Suicides	0.040	0.159 ***			4.69	<.0001
Accidents motor vehicles	0.069	0.191 ***	0.054	-0.126 **	3.77	<.0001
Other accidents						n.s.
Other non-natural	-0.197	0.472 ***	-0.468 *	0.094	2.00	0.0920

6 Cohort and Age Effects

The last chapters provide ample evidence that life span and month of birth are related. Two questions arise: first, does this influence also exist for the ages 1 to 49? Second, is the magnitude of the excess mortality of the spring-born similar in all age groups? There are at least two different hypotheses concerning these questions. A first hypothesis is that the excess mortality of the spring-born is age-specific because of the changing distribution of causes of death with age. The analysis of the US death certificates presented in Chapter 5 shows that the month-of-birth effect exists in all major groups of causes of death and that its magnitude differs, however. The excess mortality of the spring-born is largest for heart disease and considerably smaller for the group of malignant neoplasms. Thus, the month-of-birth pattern might be comparatively small at younger ages and increase its magnitude at middle and old ages when heart disease becomes an important cause of death. A second hypothesis is that the differences decrease with age due to mortality selection. Among both the spring- and the autumn-born, the frailer will die first, resulting into an ever more homogeneous and "robust" population at higher ages. In practice, it is most probably the case that both forces work simultaneously.

In trying to test these two hypotheses one has to overcome two problems. The first problem is that longitudinal data are needed to identify age-specific effects; the second problem is that datasets have to be large so as to allow for the estimation of age-specific mortality at ages where death rates are low.

Longitudinal data based on population registers exist in the Scandinavian countries. Most of them were started during the 1960s. In the case of Denmark, the register was established in April 1968, and it contains everybody who was alive in March 1968. This study uses the Danish register data together with a mortality follow-up until 1998. Because of the small number of deaths at younger ages in Denmark today, it is not possible to use the register for studying age-specific changes in the month-of-birth pattern for contemporary young cohorts. This is also true for other Scandinavian countries with population registers. The number of deaths are sufficient only after age 50 for studying age-specific patterns in mortality by season of birth. Thus, the Danish register data provide only a partial answer to the question of how the month-of-birth pattern changes with age.

The second longitudinal dataset used in this chapter is the Danish twin register, which contains twins born in Denmark between the years 1870 and 1930. The twins were followed from age six onwards until August 1995. For the study of age-specific month-of-birth patterns, twins have two advantages. First, the mortality of twins does not differ from the mortality of singletons once they survive their sixth birthday (Christensen et al. 2001). Second, the differences in life span by season of birth should be larger among twins than among singletons. There are two reasons for this. First, *in-utero* twins belong to the most vulnerable group because they have to share scarce resources, so seasonal differences in nutrition should therefore affect them more severely. Second, they have lower birth weights than singletons, which should make them more susceptible to infectious disease during their first years of life.

In theory both the Danish register data and the Danish twin data should allow us to calculate age-specific death rates by quarter of birth using simple life-table techniques. In practice, however, it appears that these death rates are highly volatile and do not lead to conclusive results. In order to arrive at consistent estimates of age-specific death rates, a functional form is imposed on the age-specific intensities of death.

Finally, a demographic method is applied that has been widely used in the study of genetic traits of longevity. This method allows for the use of repeated cross-sections of populations to estimate the effect of the "fixed attribute" month of birth on survival. In our search for repeated, large cross-sectional datasets, we obtained the US censuses for the years 1960, 1970, and 1980. These are the only three censuses that contain information about the quarter of birth. The census data were extracted from http://www.ipums.umn.edu/usa/index.html. The seasonal distribution of birth dates for, e.g., ages 10-19 in 1960 was then compared with the distribution for ages 20-29 in 1970 and for ages 30-39 in 1980. The changes in the distributions between the three census rounds indicate whether or not there exists a significant level of excess mortality of the spring-born.

All three datasets lead to the same conclusion, namely, that there is an excess mortality of the spring-born at all ages. The Danish register data show that the excess mortality after age 50 is lower in more recent cohorts and that it decreases with age. The excess mortality of the spring-born is larger for Danish twins than for the general Danish population. It exists at all ages and also tends to decrease with age. The US census data reveal that the mortality advantage of the autumn-born exists at all ages and that the difference in survival between the spring- and the autumn-born starts to accelerate at middle ages. However, the census data do not allow us to distinguish between cohort and age effects. It is therefore not obvious whether the increase in the excess mortality of the spring-born around age fifty is due to the age-specific onset of diseases with a strong month-of-birth pattern or due to cohorts that experienced worse conditions early in life.

6.1 Method and Data

The Danish register data and the Danish twin data are analysed by imposing a functional form on the age-specific force of mortality. In the Danish register data mortality increases from age 50 on, following a Gompertz function [6.1]

$$\mu(x) = ae^{bx} \qquad [6.1]$$

where $\mu(x)$ is the force of mortality at age x, a is the age-independent level of mortality and b the increase in mortality over age. The parameter estimates a and b are estimated by maximising the likelihood function. All individuals are distinguished according to their quarter of birth and for each quarter a separate model is estimated. The age-specific death rates of the four models are presented in the form of ratios of the first three quarters to the last quarter. These ratios are compared with the ratios of the age-specific probabilities of death for one-year age groups on the basis of simple life-table methods.

The Danish register data consist of sufficiently large numbers of exposures and deaths to distinguish between age and cohort effects. Two ten-year birth cohorts are followed over an age-span of 20 years. The birth cohort April 1908 to March 1918 enters the study period between the ages of 50 and 59 and 11 months. They are followed from age 60 to age 79 in order, theoretically, to allow each member of the cohort to reach each age. The second cohort is aged 60 to 69, 11 months at the 1968-baseline and is followed from age 70 to age 89. This specification allows us to study age-specific death rates at ages 70 to 79 for both cohorts.

The Danish twin data include all twins from the age of six onwards. In contrast to older ages, the force of mortality does not follow a simple parametric form between ages 6 and 100. In the twin data, mortality decreases between the ages of 6 and 15. It then increases sharply until age 20, stagnates until age 40, and increases exponentially thereafter. A possible solution is to estimate four mortality functions which are joined together at ages 15, 20, and 40.

The following model [6.2] is specified

$$\ln \mu_{ij}(x,U_j) = \alpha_0 + \alpha' y(x) + \beta' C_j + \delta I_j + U_j \qquad [6.2]$$

in which $\ln \mu_{ij}(x,U_j)$ denotes the logarithm of the force of mortality at age x for individual i of twin pair j, and $\alpha_0+\alpha' y(x)$ the age-dependent baseline hazard. Let C_j stand for the ten-year birth cohorts and their parameters β', I_j for an indicator variable that indicates the half-year people are born in (I_j=0 if they are born in the first half-year, 1 otherwise), and δ for the parameter.

Since the large amount of variability in mortality under age 20 leads to unstable estimates of the age-independent mortality level α_0, the analysis is restricted to ages 20 and above. The baseline hazards $y(x)$ is a vector of piecewise-linear spline transformations of age x

$$y(x) = \begin{pmatrix} (\min[x, \upsilon_1]), \\ (\max[0, \min[x - \upsilon_1, \upsilon_2 - \upsilon_1]]) \\ (\max[0, x - \upsilon_2]) \end{pmatrix} \qquad [6.3]$$

with the nodes υ_1 and υ_2 of the splines set at ages 40 and 60. Thus, three slopes of mortality are estimated. The first slope estimates the yearly percentage increase in mortality between ages 20 and 40. The second one estimates this for the ages 40 to 60, and the third is for ages over 60.

Since the life spans of twins are not independent, a random variable U_j is included which accounts for unobserved shared heterogeneity among twins. The random variable U_j is assumed to follow a normal distribution with mean zero and variance σ^2. Separate models are estimated for the two sexes.

In order to release the model from the proportionality assumption concerning the baseline hazard for people born in a specific half-year, model 6.4 is specified.

$$\ln \mu_{ij}(x,U_j) = z_i(\alpha_{01} + \alpha_1' y_1(x)) + (1 - z_i)(\alpha_{02} + \alpha_2' y_2(x)) + \beta' C_j + U_j \qquad [6.4]$$

In Equation 6.4 $\ln \mu_{ij}(x, U_j)$ denotes the logarithm of the force of mortality at age x for individual i of twin pair j, $\alpha_{01}+\alpha_1'y_1(x)$ and $\alpha_{02}+\alpha_2'y_2(x)$ the age-dependent baseline hazards for people born in the first and second half of the year, and z_i an indicator variable for the first or second half-year of birth (z_i=1 if born in the second half-year, z_i=0 if born in the first half-year).

Specifying two baseline functions overcomes the proportionality assumption and allows for the estimation of two separate age-independent levels of mortality (α_{01} and α_{02}) and of separate age-dependent slopes of mortality (α_1' and α_2'). The baseline hazards $y_1(x)$ and $y_2(x)$ as well as the terms C_j, β', and U_j are defined similarly to Equation 6.2. Separate models are estimated for the two sexes. All estimations are performed in aML vers. 2.

In order to compare the pattern in life expectancy by season of birth between twins and the general Danish population, life tables are calculated by week of birth. For each week of birth the life expectancy is estimated. The 52 age values are smoothed by estimating the cosine function $age = a_0 + a_1 * \cos(t - a_{11})$ with t= *week of birth*/$52*2\pi$.

To estimate differences in survival according to the quarter of birth on the basis of the US censuses for the year 1960, 1970, and 1980, a method called Survival-Attributes Assay (Vaupel 1991, Christensen et al. 2001) is applied. This method uses cross-sectional data on "fixed-attributes" to estimate the effect of a fixed trait on survival.

The Survival-Attributes Assay can be demonstrated on a simple example. Let N_{50} be the number of people at age 50. Let p_{50} be the proportion of 50-year-olds who have some fixed attribute such as the season of birth. Let p_{70} be the proportion at age 70. Let s be the conditional survival probability from age 50 to age 70 for the individuals who have the fixed attribute. Let S be the conditional survival probability from age 50 to 70 for the entire cohort.

Then, because

$$p_{50} N_{50}\, s = N_{50}\, S\, p_{70}, \qquad [6.5]$$

it follows that

$$s = S\, p_{70} / p_{50}. \qquad [6.6]$$

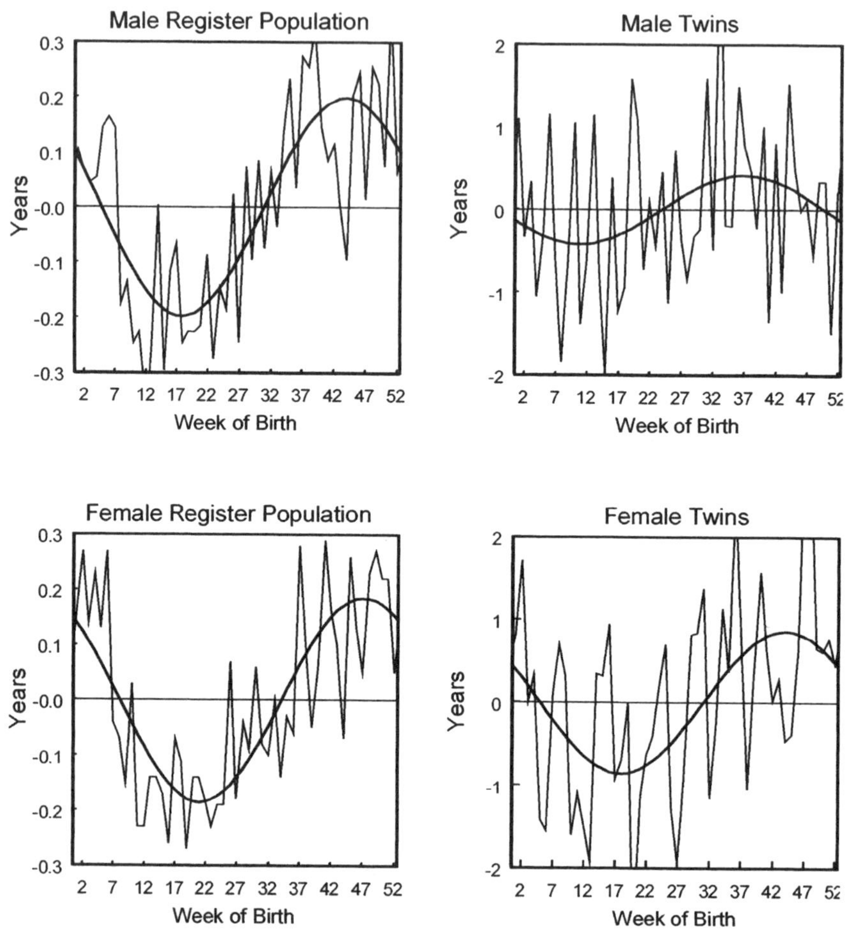

Figure 6.1 Differences in life expectancy by week of birth in the Danish register population (e^{50}) and the Danish twin register (e^{6}). The fitted lines are estimated by sinusoidal functions with a period of 52 weeks.

Thus, the relative risk of surviving from age 50 to age 70 for people born in a specific quarter is the ratio of their observed proportions in the two cross-sections. In this study we calculate the proportion of the population within ten-year age groups that is born in a certain quarter of the year. These age groups are then followed over the three census rounds.

This is a very simple and useful method but it relies on some crucial assumptions. The most important of these assumptions is that the 70-year-olds in the third cross-section were similar to the 50-year-olds in the first cross-section. The three US census rounds used in this study, however, allow us to follow a ten-year cohort on an aggregate level over a period of 20 years. The 70-year-olds in 1980 are the survivors of the 60-year-olds in 1970 and the 50-year-olds in 1960. In other words, the change in the proportion of the fixed attribute over an age range of twenty years is solely due to age effects and cohort effects do not confound it. Cohort effects confound comparisons of ages further apart than 20 years, however.

The three US census rounds 1960, 1970, and 1980 are used because these are the only census rounds that include information about the quarter of birth. Data are extracted from the "Public Use Microdata Samples", which are accessible under http://www.ipums.umn.edu/usa/index.html. The extract is restricted to the native-born white and black US population aged 0 to 100. For whites, this gives a sample size of 1,490,444 in 1960, 1,672,107 in 1970, and 1,812,839 in 1980. For blacks, we have 187,849 observations for the year 1960, 220,670 for 1970, and 256,447 for 1980.

The Danish population register was described in the second chapter of this manuscript. The Danish twin registry was established in 1954 as the first nationwide twin registry in the world. It includes all twin pairs born in Denmark between 1870 and 1910 and all same-sex pairs born between 1911 and 1930. The birth registers from all 2,200 parishes of the relevant calendar years were manually scrutinized to identify all twin births. Through regional population registers (in operation since 1924) and other public sources, a search was made for the twins, or whenever needed, their closest relatives. For twins who died or emigrated at an early age, it was impossible to obtain reliable data to be used in zygosity classification. Consequently, pairs were not followed up if one or both partners died or emigrated before the age of six. In this study 12,530 female and 11,237 male twins are followed until August 1995, and for each twin the exact age at death or at censoring is known.

6.2 Results

6.2.1 Comparison of the Month-of-Birth Pattern in the Danish Register Data and the Danish Twin Data

In both populations, life expectancy decreases sharply for those born in the first quarter. It reaches a trough for people born in the second quarter and then increases sharply for the summer-born. Life expectancy peaks for people born in the autumn.

The fit of the sinusoidal functions is significant for both sexes (p=0.001) and both registers. The striking difference between the pattern of the twins and that of the general population lies in the amplitude. In the female register population the amplitude is 0.19 years and for twins it is 0.85 years; for males the difference is 0.20 years and 0.42 years, respectively. Thus, the peak-to-trough difference among the female twins is about four times as large as among the register population. In interpreting this difference one has to keep in mind that, for twins, it reflects the difference in life expectancy from age six onwards while, for the general Danish population, it refers to differences in remaining life expectancy at age 50.

6.2.2 Age-Specific Mortality at Ages 50+ in the Danish Register Data

Figures 6.2a,b contain the parameter estimates *a* and *b* of the Gompertz models for the two cohorts 1889 to 1908 and 1909 to 1918, which are the basis for the mortality ratios displayed in Figure 6.3. As previously discussed, in the Gompertz-model the parameter *a* depicts the age-independent mortality level and *b* the increase in mortality over age. Excess mortality of the spring-born as compared to the autumn-born implies a significantly higher level for parameter *a* for the spring-born. If the excess mortality is age-dependent then parameter *b* should also differ significantly.

Significant differences in both the level of and increase in mortality between the spring- and the autumn-born exist in the older female cohort, although the increase in mortality is only of borderline significance. The level of mortality is lower among the autumn-born while the increase in mortality with age is higher. This implies that, with advancing age, mortality increases more rapidly among the autumn-born and that the excess mortality of the spring-born becomes lower. This is also true for males and in the younger female cohort, although the results are not significant.

Figure 6.3 shows the ratio of the age-specific death rates for people born in the first three quarters of the year to the death rates of those born in the fourth quarter. The smooth lines are based on the mortality estimates from the Gompertz models, the fluctuating lines on the life-table estimates. The solid lines pertain to the birth cohort 1909 to 1918, the dashed lines to the birth cohort 1898 to 1909. For men there is little evidence of age-specific mortality patterns after age 60, with the exception that the excess mortality of those born in the first half of the year decreases between ages 70 and 89. Among females this differs insofar as the excess mortality tends to increase between ages 60 and 70 and then starts to decrease at later ages.

There is still a decrease in excess mortality for the higher ages. There is a strong cohort effect among women and a weak effect among men. In the overlapping age groups 69 to 79 the excess mortality of the spring-born is much larger in the older female cohort than in the younger one. Although

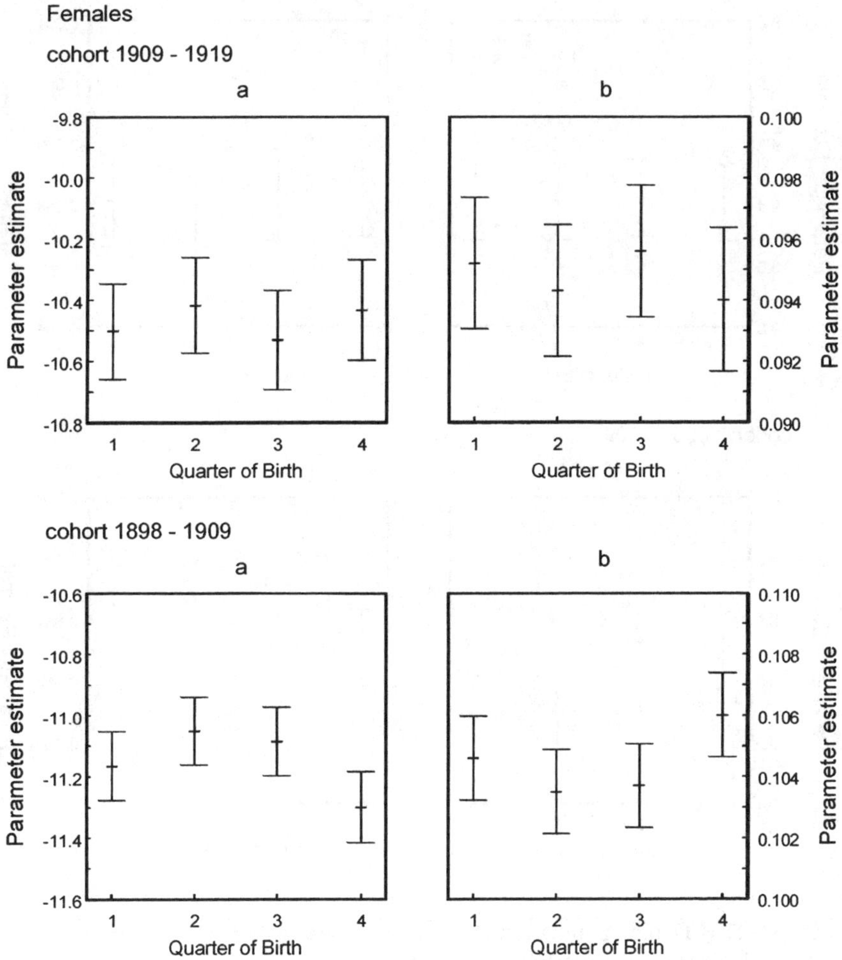

Figure 6.2.A Danish population register: parameter estimates *a* (level of mortality) and *b* (increase in mortality with age) of the Gompertz model specified in Equation 6.1 and 95% confidence intervals by quarter of birth for the two birth cohorts April 1909 – March 1918 and April 1898 – March 1908, females.

there is a similar tendency for all cohorts and both sexes one has to keep in mind that, on the basis of the Gompertz model, the differences are only significant for the older female cohort.

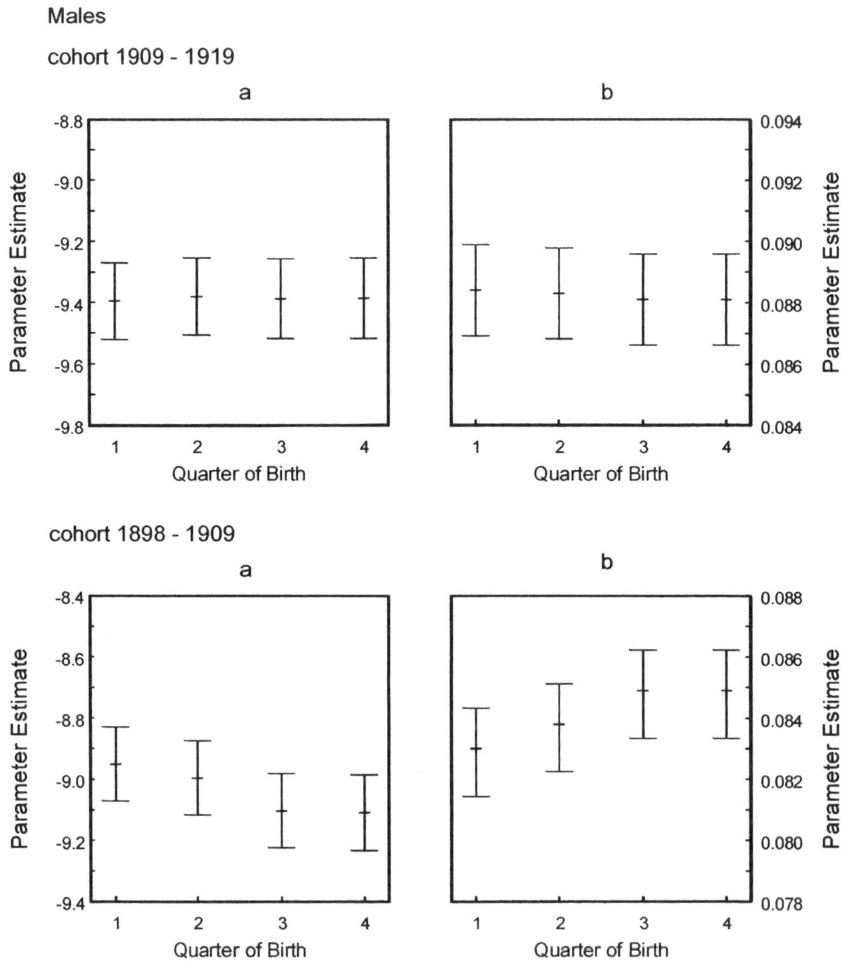

Figure 6.2.B Danish population register: parameter estimates *a* (level of mortality) and *b* (increase in mortality with age) of the Gompertz model specified in Equation 6. 1 and 95% confidence intervals by quarter of birth for the two birth cohorts April 1909 – March 1918 and April 1898 – March 1908, males.

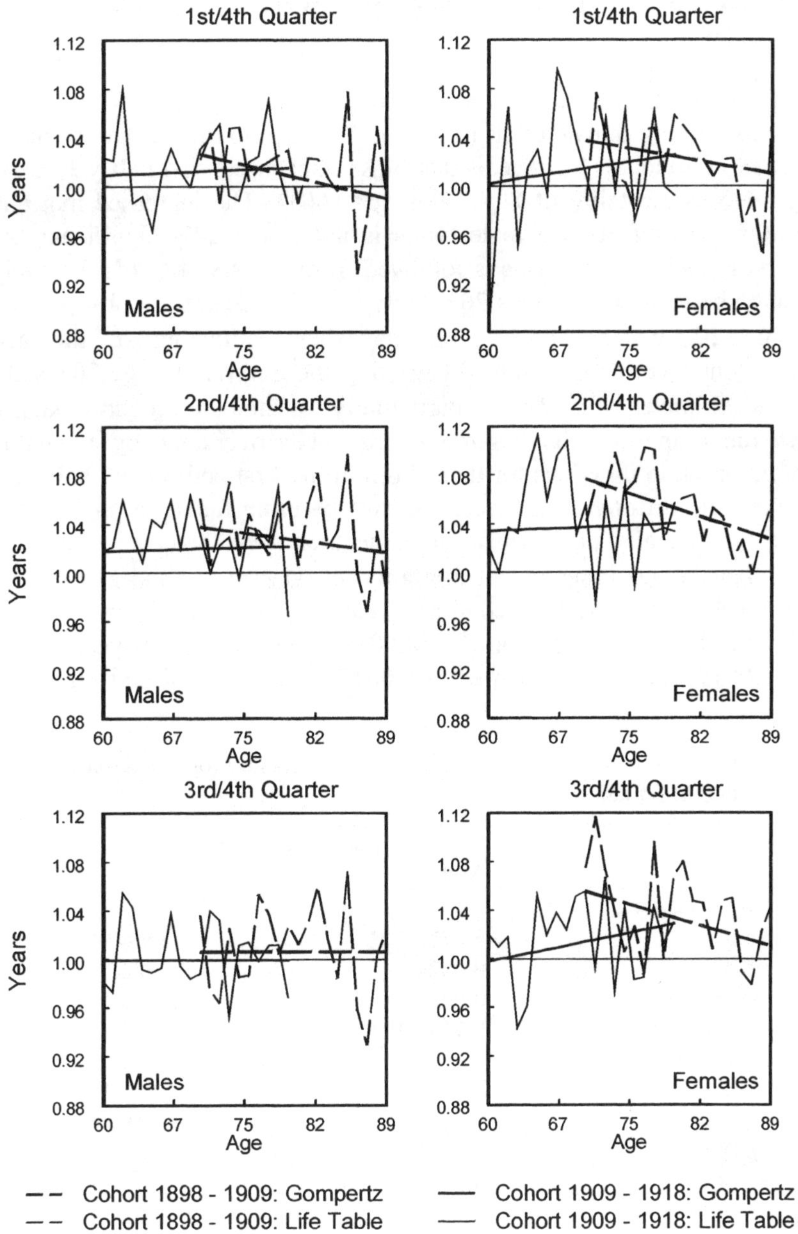

Figure 6.3. Danish population register: rate ratios by quarter of birth for the two birth cohorts April 1909 – March 1918 and April 1898 – March 1908.

6.2.3 Age-Specific Mortality at Ages 20+ in the Danish Twin Data

Significant differences in the force of mortality after age 20 by half-year of birth exist among female twins (Table 6.1). The proportional hazard model specified in Equation 6.2 reveals that those born in the first half-year experience excess mortality of 8 per cent ($p<0.001$). The increased mortality risk of 4 per cent among male twins is not statistically significant. Although the mortality of twins is followed from age six onwards the analysis has to be restricted to ages 20+. Only very few deaths are observed under age 20, which causes problems when estimating α, the age-independent level of mortality. Restricting the analysis to age 20+ stabilizes the estimates of α. This is particularly true for the second model, in which the proportionality assumption has been overcome by estimating separate baseline functions for those born in the first and second half-year. Although the differences between the two baseline functions in Equation 6.4 are statistically not significant, a general trend emerges. Table 6.2 shows that the age-independent mortality levels $exp(\alpha_{01})$ and $exp(\alpha_{02})$ are lower for those born in the second-half year (males first half-year: $e^{-6.17}=0.0021$, second half-year $e^{-6.25}=0.0019$; females first half-year: $e^{-6.26}=0.0019$, second half-year $e^{-6.41}=0.0017$). Figure 6.4 displays the loga-

Table 6.1. Danish twin data: parameter estimates of the proportinal hazard model specified in Equation 6.2

	Parameter estimates	
	Males	Females
Baseline hazard		
constant	-6.2282 ***	-6.3696 ***
slope 20-40	0.0020	-0.0002
slope 40-60	0.0881 ***	0.0698 ***
slope 60+	0.1170 ***	0.1157 ***
Birth cohort		
1870-1879	-0.0354	0.3532 ***
1880-1889	0.1167 **	0.2919 ***
1890-1899	0.0320	0.1725 ***
1900-1909	0.0503	0.0013
1910-1919	-0.0352	-0.0922 *
Born in first half-year	0.0443	0.0802 ***
σ^2	0.6321 ***	0.5522 ***
ln-L	-81186.0	-81397.3

*** p=0.001, **p=0.05, *p=0.1

rithmic hazards and Figure 6.4b the ratios of the hazards of the first and second half-year born. The latter reflects the excess mortality of those born in the first half of the year, which varies by sex. Among males, the excess mortality of the first half-year born increases between ages 20 and 40; among females, it decreases. An opposite trend exists between ages 40 and 60: excess mortality decreases among males and increases among females. From age 60 onwards the differences start to decline for both sexes. In interpreting this result one has to keep in mind that the maxima and minima in the trajectories of the hazard ratios in Figure 6.4b are determined by the positioning of the knots at ages 40 and 60. The knots are positioned such that they reflect changing points in the increase of the force of mortality with age. They do not necessarily reflect the ages where the difference between the first and second half-year born is largest or smallest.

Table 6.2. Danish twin data: parameter estimates of the hazard model with two baseline functions specified in Equation 6.3.

	Parameter estimates	
	Males	Females
Baseline hazard born in 1st half-year		
Constant α_{01}	-6.1696 ***	-6.2545 ***
Slope 20–40	0.0035	-0.0039
Slope 40–60	0.0858 ***	0.0734 ***
Slope 60+	0.1166 ***	0.1135 ***
Baseline hazard born in 2nd half-year		
Constant α_{02}	-6.2447 ***	-6.4126 ***
Slope 20–40	0.0001	0.0041
Slope 40–60	0.091 ***	0.0658 ***
Slope 60+	0.1174 ***	0.1179 ***
Birth cohort		
1870–1879	-0.0379	0.3545 ***
1880–1889	0.1164 **	0.2924 ***
1890–1899	0.0317	0.1745 ***
1900–1909	0.0507	0.0027
1910–1919	-0.0354	-0.0902 *
σ^2	0.6323 ***	0.5524 ***
Ln-L	-81185.3	-81395.9

***p=0.001, **p=0.05, *p=0.1

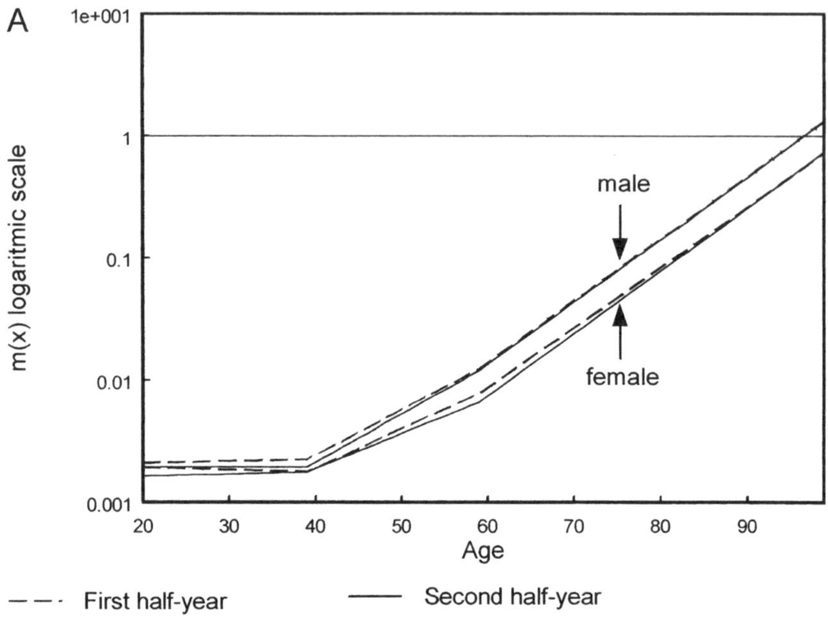

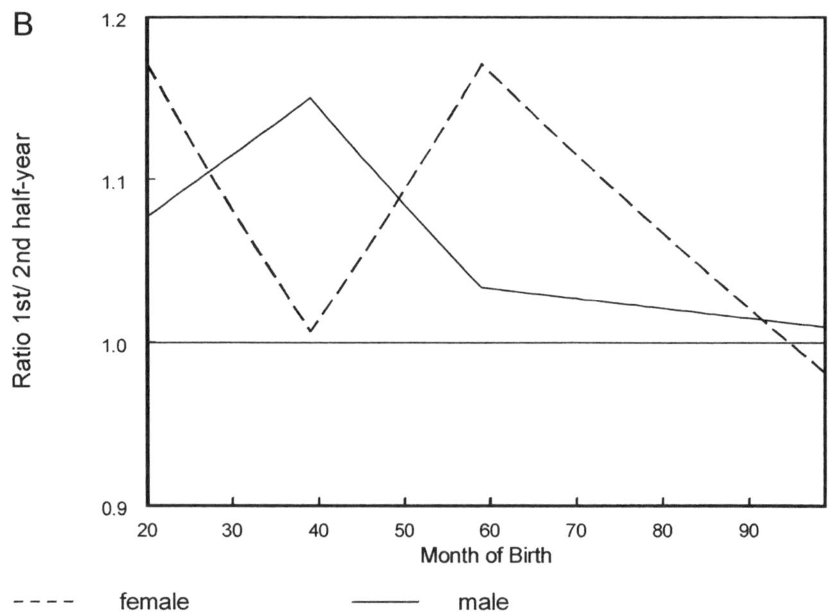

Figure 6.4. Danish twin data: (A) hazards for those born in the first and second half-year of the hazard model specified in Equation 6.4 and (B) ratios of the hazards of those born in the first/second half-year.

6.2.4 US Census Data

The eight trajectories in each square of Figures 6.5.A and 6.5.B depict the change in the distribution of people born in a specific month as they age. For example, the first trajectory shows the percentage of people aged 0-9 in 1960, 10-19 in 1970, and 20-29 in 1980; the second trajectory shows the percentage of people aged 10-19 in 1960, 20-29 in 1970, and 30-39 in 1980. The differences in the starting values of the trajectories for a particular age group are probably random effects due to census sampling.

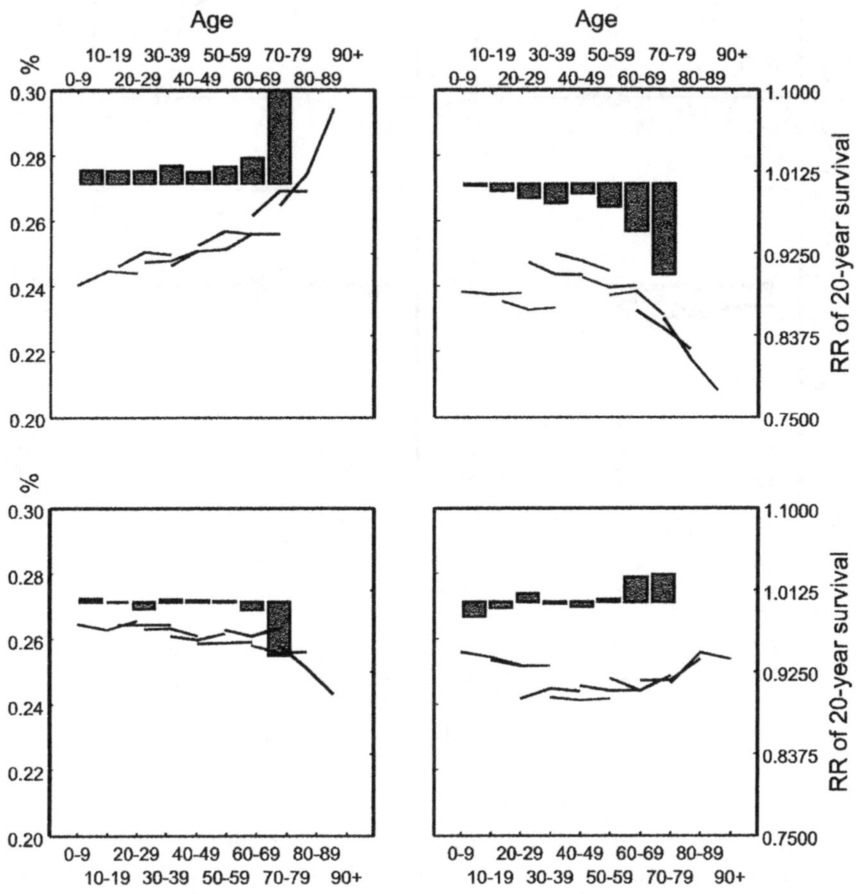

Figure 6.5.A US census data: changes in the proportion of people born in a specific month (lines) and relative risks of 20-year survival (bars) for US whites, based on the census data for the years 1960, 1970, and 1980; females.

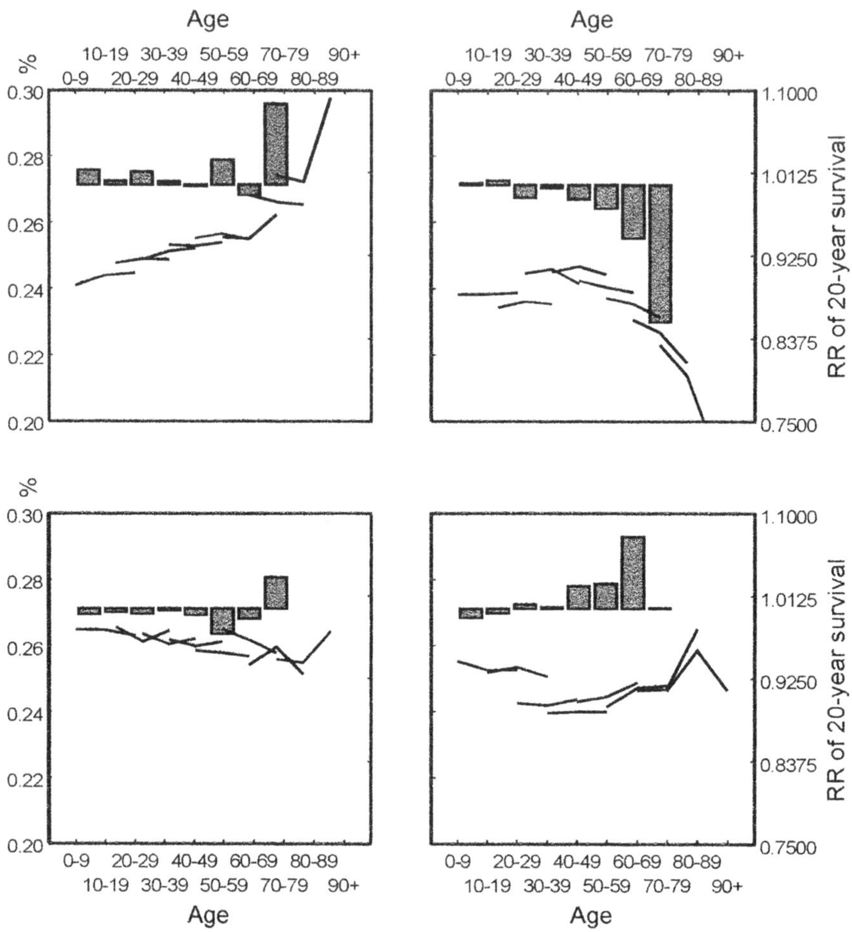

Figure 6.5.B US census data: changes in the proportion of people born in a specific month (lines) and relative risks of 20-year survival (bars) for US whites, based on the census data for the years 1960, 1970, and 1980; males.

Among whites the proportion of people who are born in the first and the fourth quarters increases continuously with age; the proportion of those born in the second and third quarters decreases continuously. Based on the Chi-square test statistic, the changes in the proportion are significant starting from the age group 30 to 39. The mortality disadvantage of the spring-born seems to be particularly large from age 60 onwards, however.

The decrease between ages 0 and 39 in the proportion of people born in the fourth quarter is highly significant (p=0.001).

An rather similar age pattern exists for blacks (results not shown): a continuous increase in the proportion of those born in the first quarter of the year; a decrease in the proportions of those born in the second and the third quarters. The pattern of those born in the fourth quarter is less consistent, suggesting both an increase in the proportion during middle ages and a decrease at older ages. For blacks, the changes in the distribution tend to be not significant, mainly due to the smaller numbers of observations, especially for the oldest age groups.

The US results from the comparison of the three census rounds are consistent with the results from the US death data. An increasing proportion with age of those born in the first and the fourth quarters is consistent with a higher mean age at death.

The bars in Figure 6.5a and 6.5b show the relative risks of the 20-year survival probabilities conditional on age for people born in a specific quarter compared to the average population. At younger ages the mortality advantage of the autumn-born and the disadvantage of the spring-born is minor. For males it is a maximum of one per cent over an age range of 20 years. In other words, up to the age of 40 the conditional survival probability of surviving the next 20 years is about one per cent higher for the autumn-born than for the average population; it is one per cent lower for the spring-born. Differences in the 20-year survival start to accelerate from the age group 40-49 onwards, when those born in the fourth quarter have a higher chance of 2.4% to survive the next 20 years; this advantage increases to 7.8 % for ages 60-69. The disadvantage in the 20-year survival of people born in the second-quarter starts at ages 50-59 and is about minus 2.8 per cent. It increases to minus 6.6 per cent for the age group 60-69. Similar trends emerge for women, which are more consistent, however.

6.3 Discussion and Conclusion

A look at all three populations leads us to the same conclusion: the mortality advantage of the autumn-born and the disadvantage of the spring-born exist in all age groups, but its magnitude is age-dependent. Both among the Danish twins and the general Danish population, the excess mortality of those born in the first half-year decreases in the highest age groups. In both populations this decrease is stronger among females than among males. The decreasing excess mortality at the highest ages is consistent with the effect of mortality selection. Among both the spring- and the autumn-born only the strongest survive up to the highest ages and differences according to the month of birth start to fade out.

The magnitude of the excess mortality differs also at younger ages. The results from the twin data suggest that, among men, the excess mortality increases between age 20 and age 40 and slowly declines thereafter. The slow decline in the excess mortality above 40 is supported by the results from the general Danish population. Among female twins excess mortality decreases between age 20 and 40, increases between 40 and 60 and declined rapidly thereafter. Again, the general Danish population trend supports this finding, where the peak of excess mortality appears to be in the mid-60s and is followed by a rapid decline.

Most interestingly, among female twins the excess mortality of those born in the first half-year disappears at middle ages. Pre-menopausal breast cancer may be one explanation. A Swedish study that compared the birth distribution of 115,670 women with breast cancer to the total number of live births in their birth cohorts found different month-of-birth patterns for women with pre-menopausal and post-menopausal breast cancer. The risk of pre-menopausal breast cancer was increased for women born in the autumn, and the risk of postmenopausal breast cancer was increased for those born in late spring (Yuen et al. 1994).

A comparison of the week-of-birth pattern in life expectancy among the general Danish population and the Danish twins shows a high degree of correlation, particularly among females. The male twin pattern, however, diverges to a larger extent from the pattern of the male general population, although the fitted sinusoidal function is significant. In particular, male twins born in the fourth quarter do not experience a mortality advantage similar to that of males in the Danish population.

Looking at the Danish register data, the difference in remaining life expectancy at age 50 between the autumn- and the spring-born is 0.3 years; among the Danish twins it ranges from 0.58 years in the youngest cohort (1910-1930) to 0.88 years in the cohort 1890-1909. In other words, for both sexes together the difference is at least two to three times higher among twins than among the general population. For females, the difference is about four times as larger in the twins than in the register population. Larger differences among twins than among the general population are in accordance with the hypothesis that seasonal changes in nutrition and infectious disease affect the foetus *in-utero* and the infant in the first year of life. Twins have to share scarce resources in-utero, they usually have lower birth weights than singletons, and they have a higher risk of mortality during their first years of life.

In Denmark the extent of the peak-to-trough difference in life span changes over cohorts. At similar ages they are larger in the older cohort

(1898-1908) than in the younger cohort (1909-1918). This decrease was first described in Doblhammer & Vaupel (2001). The age-specific analysis now provides further evidence that the extent of the season-of-birth pattern in the life span decreases over the time period from 1898 to 1918, when considerable improvements in nutrition took place and public health measures reduced the risk of infectious diseases. The decrease over birth cohorts appears to be stronger among females than among males.

The age-specific analysis on the basis of the three US censuses further supports the finding that the mortality advantage of those born in the second half of the year exists at all ages. In contrast to the Danish data, however, the mortality disadvantage of the spring-born starts to accelerate from age 40-50 onwards. The most probable explanation is that this increase reflects a cohort effect rather than an age effect since the US data do not allow one to distinguish between age and cohort effects for ages more than twenty years apart. Particularly the oldest age groups were born at a time before dietary patterns started to change and to resemble contemporary dietary patterns (Levenstein 1988, 1993). They were also born before improvements in infrastructure and the introduction of public health measures that significantly reduced infant mortality – particularly mortality caused by infectious disease (Preston & Haines 1991).

In all three datasets, namely, the Danish twins, the Danish register population, and the US censuses, the extent of the age-specific month-of-birth patterns seems to be sex-specific. First, the decrease in the excess mortality at advanced ages is more pronounced among females than males; this appears to be true also for the decrease over cohorts. Second, the twenty-year survival risk in the US censuses appears to be more consistent for females than males. Third, among females the excess mortality of those born in the first-half year reaches its maximum at a later age (60s) than among males (40s). Sex-specific causes of death may be responsible for the finding that, at middle ages, the mortality advantage of the autumn-born disappears among female twins. This result is consistent with a study conducted by Gavrilov et al. (2002), which reports sex-specific differences in the month-of-birth pattern among the European nobility. In particular, they find a more consistent pattern among females with higher excess mortality than among males.

The month-of-birth pattern in adult mortality exists at all ages. Over the whole life course those born in the first part of the year experience higher mortality than those born in the second part. The month-of-birth pattern is age-specific for two reasons: first, different causes of death are important at different ages and the magnitude of the excess mortality depends on the cause of death. Second, mortality selection results in a more homogenous group at higher ages because among people born both in the first and second half-year, only the most robust individuals survive, which means that differences are reduced.

All three datasets provide evidence that the month-of-birth pattern decreases in younger cohorts. This finding raises the question whether the differences still exist for people born today or whether the differences in life span by month of birth are merely of historical interest. An answer to this question will be provided in Chapter 8.

7 The Month-of-Birth Patterns of Migrants and Farmers

The second chapter provides evidence that the month-of-birth pattern is tied to the seasons of the years by comparing differences in the life span of people born in the Northern and the Southern Hemisphere. This idea is taken one step further in the analysis of the month-of-birth pattern of migrants. Two groups of migrants are explored. The first group moved from the Northern to the Southern hemisphere – more precisely, it consists of people who were born in Britain and who died in Australia. The second group consists of internal US migrants who moved between different regions of the US and, thus, died in a region other than the one in which they were born. The underlying idea is that migrants should be subject to the month-of-birth pattern of their region of birth rather than that of their region of residence. This chapter provides evidence that this conjecture is indeed true. However, the month-of-birth pattern of people who migrated from the South to the North of the US does not always lead to consistent results.

The second part of the chapter explores whether unobserved social differences exist in the month-of-birth pattern, and whether these unobserved social differences may explain the inconsistencies in the South to North migration patterns. The only socioeconomic information on US death certificates is education and usual occupation (for some states only). Education probably does not capture the main socioeconomic differences that affect early-life circumstances in the South at the beginning of the 20th century. At that time the South was still an agricultural society, with the majority of the population born in rural areas. The educational system was much less developed than in the industrialized North – a large majority of the population had only basic education. On the other hand, the rural population was far from being a homogenous group, since the social class relations in the South were structured along the lines of land ownership. In this social structure those who did not own land were particularly disadvantaged.

The second piece of socioeconomic information on the death certificates is "usual occupation", which is generally difficult to relate to early-life circumstances. There is one exception, namely, the two occupations farmer and farm worker. Farmers and farm workers are generally born in rural areas, and Preston et al. (1998) showed that a rural place of birth and, particularly, a farming background increases the old-age survival of African-Americans. Thus, the conjecture is that the month-of-birth pattern of farmers and farm workers should differ from that of the remaining population, since their early-life circumstances were different in terms of nutrition and infectious disease environment. Furthermore, farm workers in the South who did not own land were at a particular disadvantage, and one can therefore expect that fluctuations in the month-of-birth pattern are larger for farm workers than for farmers.

If differences in the month-of-birth patterns of the urban and rural-born populations exist, and if there are differences in the pattern among the rural population associated with the ownership of land, then this might explain the inconsistencies in the month-of-birth pattern of migrants from the southern to the northern US. The question of whether the migrants of a particular region of the US came primarily from urban or rural areas and whether or not they owned land might have a significant influence on the month-of-birth pattern in mean age at death.

7.1 Data and Method

7.1.1 Migrants

The Australian death data provide information about the age at immigration and the country of origin. They contain a total of 43,074 British-born immigrants to Australia who died at ages 50+ with known age at immigration.

The United States consists of nine geographical regions according to the classification given in the death records. All decedents whose state of residence differed from their state of birth are treated as migrants (see Tables 7.1 and 7.2). The largest migration flow among US born decedents is from the North to the South: 438,061 decedents were born in the Middle Atlantic region and lived in the South Atlantic region at the time of their death; the second largest is from North to West: 330,396 migrants were born in the West North Central region and moved to the Pacific region; the third largest flow is from the South to the North: 283,129 decedents born in the East South Central region moved to the East North Central region.

In order to test the hypothesis that the age at death of migrants follows the month-of birth pattern of their birth region rather than of their residence region, regions with substantially different month-of-birth patterns are needed. This is true for the South and the North of the US.

Equation 7.1 is used to estimate the simultaneous effects of the characteristics of the decedents on their age at death.

$$x_j = \alpha_0 + \beta' Y_j + \gamma' I_j + u_j . \qquad [7.1]$$

Let x_j be age at death, Y_j the matrix of the indicator variables month of birth, sex, education, migrant and marital status, and I_j the matrix of the two-way interactions between the variables; α_0, β', and γ' are the parameter estimates. The error term u_j follows a normal distribution with mean zero and variance σ^2. The variable migrant indicates whether a person died in the birth region or has moved to another region. Of particularly interest is the two-way interaction between the variables month-of-birth and migrant because the value of the F-statistics indicates whether the month-of-birth pattern of the migrants differs significantly from that of their birth or residence regions.

For each migration flow, two models are estimated. The first model compares the non-migrants of the birth region with those who migrated, e.g. decedents born in the Middle Atlantic region who remained in their birth state with those who moved to the South Atlantic region. The second model compares the migrants with the non-migrants of the residence region, e.g. the migrants from Middle Atlantic to South Atlantic with those who were born and who died in South Atlantic.

7.1.2 Farmers, Non-Farmers and Farm Workers

Only a subset of the US states reports the usual occupation in their death records - in the North these are the states Indiana, Kansas, Maine, New Hampshire, New Jersey, Ohio, Rhode Island, Vermont and Wisconsin. In the South, Georgia, Kentucky, North Carolina, Oklahoma, South Carolina and West Virginia provide the information and in the Mountain region, Colorado, Idaho, Nevada, New Mexico and Utah do so.

All people who died in their birth regions were extracted. In the North and the Mountain region the analysis is restricted to white males only because of the different month-of-birth pattern for blacks and whites. For the South the large number of black farmers allows for the estimation of a separate model for blacks. Females were excluded from the analysis because female farmers may be classified as housewives.

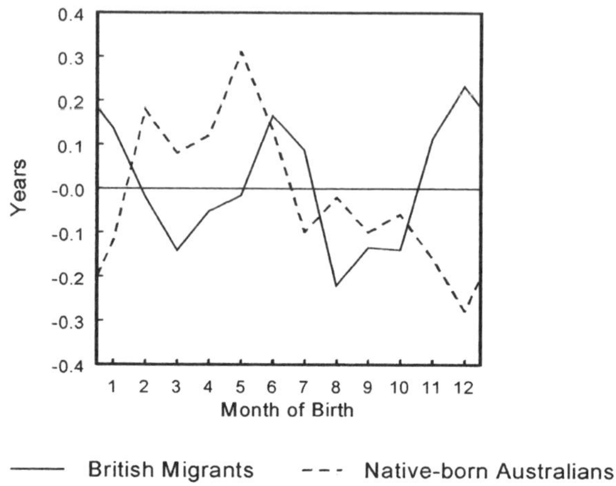

Figure 7.1. Deviation in the mean age at death by month of birth from average age at death of British immigrants and native-born Australians.

Occupations in the death records are classified according to the "Instruction Manual of the Alphabetical Index of Industries and Occupations, 1998". The 1990 census code for farmer is 473, and the code for farm worker is 479. These two groups are compared to the residual group "nonfarmers"; death records with unknown occupation are excluded.

In the North death records report farmer or farm worker as the usual occupation for 53,143 white males. That is about 9 per cent of the death records that contain information about the usual occupation. In the South there are 54,328 white male farmers (12 per cent) and in the mountain region 10,516 (14 per cent). Among blacks in the South 15,052 (13 per cent) are farmers.

A regression model similar to Equation [7.1] is estimated. The model includes the indicator variables region, marital status, education, month of birth and farmer, and the two-way interactions between the variables month of birth and farmer, and education and farmer. The variable farmer distinguishes farmers from all others, who are called non-farmers.

A second model with a similar specification is estimated which distinguishes between farmers and farm workers. Since farmers and farm workers have similar educational levels, education is not included in the latter model. Among whites there are 51,160 farmers and 1,983 farm workers in the North; 52,572 farmers and 1,576 farm workers in the South and 9,555 farmers and 961 farm workers in the Mountain region. Among blacks in the South there are 11,736 farmers and 3,316 farm workers.

7.2 Results

7.2.1 British Migration to Australia

The comparison of the month-of-birth pattern of British immigrants with the pattern of people born in Australia reveals a striking difference (Figure 7.1). The peak in mean age at death of British immigrants to Australia born between November and January is 0.36 years higher than the trough of their Australian-born counterparts (one sided t-test, $p<0.001$). The life span of British immigrants born between March and May is 0.26 years lower than that of native Australians (one sided t-test: $p<0.007$). However, the month-of-birth pattern of the migrants is bimodal, while the pattern for Austria, Denmark, or the total US is unimodal (see Fig. 2.1). One explanation may be that the month-of-birth pattern for the total population in Britain is bimodal, another, that migrants are not representative of the total population. They may be socially selected, and there is ample evidence that they are selected for better health status (Young 1987, Ringbäck et al. 1999, Razum et al. 1998, Swerdlow 1991, Kliewer 1992, Lechner & Mielck 1998).

Australian death data contain the amount of time that foreign-born citizens have lived in Australia before their death. In theory this information would allow for one to test whether the critical period early in life is *in-utero* or later. If the critical period is *in-utero* then the month-of-birth pattern of migrants that came during their first year of life should resemble the pattern of their birth countries overseas. Thus, the correlation with the month-of-birth pattern of native-born Australians should be negative. A positive correlation would be an indication that the critical period is the first year of life. In practice, one cannot distinguish between the period *in-utero* and the first year of life because age at immigration can only be calculated in full years. In other words, migrants who came during their first year of life may have spent a few weeks or almost a whole year in their birth country. A second problem is that only 358 British immigrants came during their first year of life. Among these migrants no significant month-of-birth pattern exists, with the exception of a particularly low mean age at death for those born in June. The correlation between the month-of-birth pattern of the migrants who came during their first year of life and the

Table 7.1. Number of deaths by region of birth and region of residence, US death records 1989-1997, whites.

White decedents	Region of Residence									
Region of birth	NE	MA	ENC	WNC	SA	ESC	WSC	M	P	TOTAL
NE	667,698	49,434	16,469	4,345	101,438	4,602	9,355	15,619	48,911	917,871
MA	81,430	1,942,430	124,060	16,551	429,699	17,058	36,256	62,723	150,729	2,860,936
ENC	13,796	47,736	1,981,961	87,133	242,174	41,024	66,930	114,219	233,541	2,828,514
WNC	5,684	14,938	118,286	1,035,654	62,914	15,037	90,713	150,613	326,102	1,819,941
SA	8,012	50,195	75,902	8,193	1,398,012	52,868	29,238	18,633	41,299	1,682,352
ESC	2,387	8,827	143,998	15,736	136,889	793,420	68,853	17,933	40,192	1,228,235
WSC	2,247	6,682	31,140	46,927	35,410	28,353	1,082,852	70,190	179,791	1483,592
M	1,596	3,603	9,885	13,357	9,607	2,240	17,087	245,472	141,450	444,297
P	2,058	4,120	6,908	6,225	10,382	2,171	10,036	34,969	492,391	569,260

NE: New England; MA: Middle Atlantic; ENC: East North Central; WNC: West North Central; SA: South Atlantic; ESC: East South Central; WSC: West South Central; M: Mountain; P: Pacific

Table 7.2. Number of deaths by region of birth and region of residence, US death records 1989-1997, blacks.

Black decedents	Region of Residence									
Region of birth	NE	MA	ENC	WNC	SA	ESC	WSC	M	P	TOTAL
NE	6,422	1,415	298	50	792	58	60	77	420	9,592
MA	1,250	69,531	3,473	235	8,362	351	477	516	2,550	86,745
ENC	334	2,869	54,100	1,524	2,484	1,185	879	811	4,537	68,723
WNC	104	779	6,054	17,499	735	358	816	1,043	4,294	31,682
SA	11,461	149,515	49,969	1,461	491,893	7,057	2,916	1,586	10,160	726,018
ESC	2,264	20,739	139,131	16,138	23,708	196,459	15,665	2,565	21,036	437,705
WSC	587	4,405	35,735	14,921	3,963	4,373	215,952	7,975	66,374	354,285
M	15	131	230	124	137	52	185	776	917	2,567
P	26	227	210	58	175	47	158	157	5,331	6,389

NE: New England; MA: Middle Atlantic; ENC: East North Central; WNC: West North Central; SA: South Atlantic; ESC: East South Central; WSC: West South Central; M: Mountain; P: Pacific native-born Australians is negative (ρ=-.209, p=.519) but not significant. The same is true for those who came during their second (ρ=-.265, p=.405, n=569) and third year of life (ρ=-.024, p=.940, n=778). If one combines those who came during the first two years of life, there exists a negative correlation of ρ=-.344 at p=.27. The month-of-birth patterns of the migrants who came after their third year of life and the native-borns are unsystematic and statistically not significant.

7.2.2 Internal US Migration

White North-to-South Migration

Part A of Figure 7.2 compares the month-of-birth pattern of the migrants from the regions *Middle Atlantic* to *South Atlantic* with the pattern of the non-migrants of their birth region (Middle Atlantic). Part B compares it with the pattern of the non-migrants of their residence region (South Atlantic). The month-of-birth pattern of the migrants does not differ significantly from the pattern of the non-migrants of the birth region; it differs significantly from the pattern of the non-migrants of the residence region. Results are similar for the two other North-South migration flows from *New England* to *South Atlantic* and from *East North Central* to *South Atlantic*.

White South-to-North Migration

The results are more complex for the South-North migration flows (Fig. 7.3). The month-of-birth pattern of migrants from *East South Central* to *East North Central* differs significantly from the pattern of their birth regions and their residence regions. Among the migrants, the difference between spring- and autumn-born is smaller than among the non-migrants of their southern birth region; it is larger than among the non-migrants of their northern residence region. The correlation between the month-of-birth pattern of migrants and of non-migrants of the birth region is larger (ρ=0.811, p=0.001) than between migrants and non-migrants of the residence region (ρ=0.734, p=0.007). The pattern of migrants from *South Atlantic* to *East North Central* differs neither from their birth nor from their residence region, and the same is true for migrants from *South Atlantic* to *Middle Atlantic*. For both migration flows, the correlation between the month-of-birth patterns is contrary to the expectation. The pattern of the

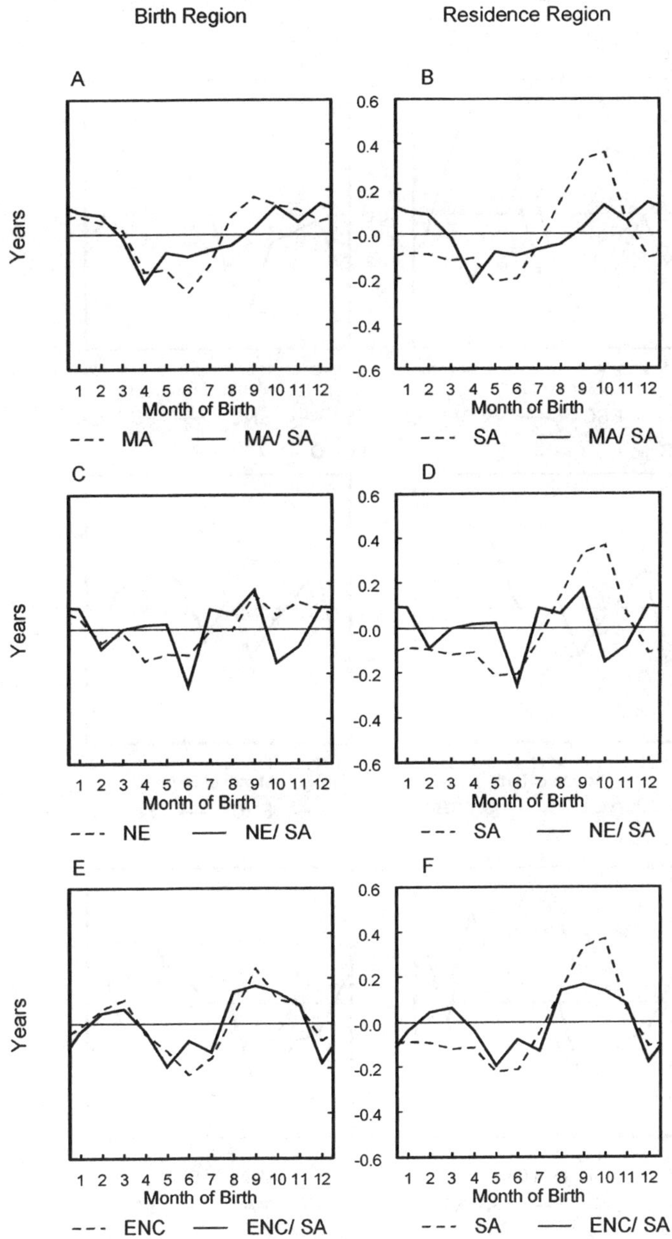

Figure 7.2. The month-of-birth pattern for white migrants from the North to the South: differences in mean age at death by month of birth from the average mean age at death at ages 50+. A,C,E) non-migrants' birth region (dashed lines), migrants (solid lines); B,D,F) non-migrants' residence region (dashed lines), migrants (solid lines).

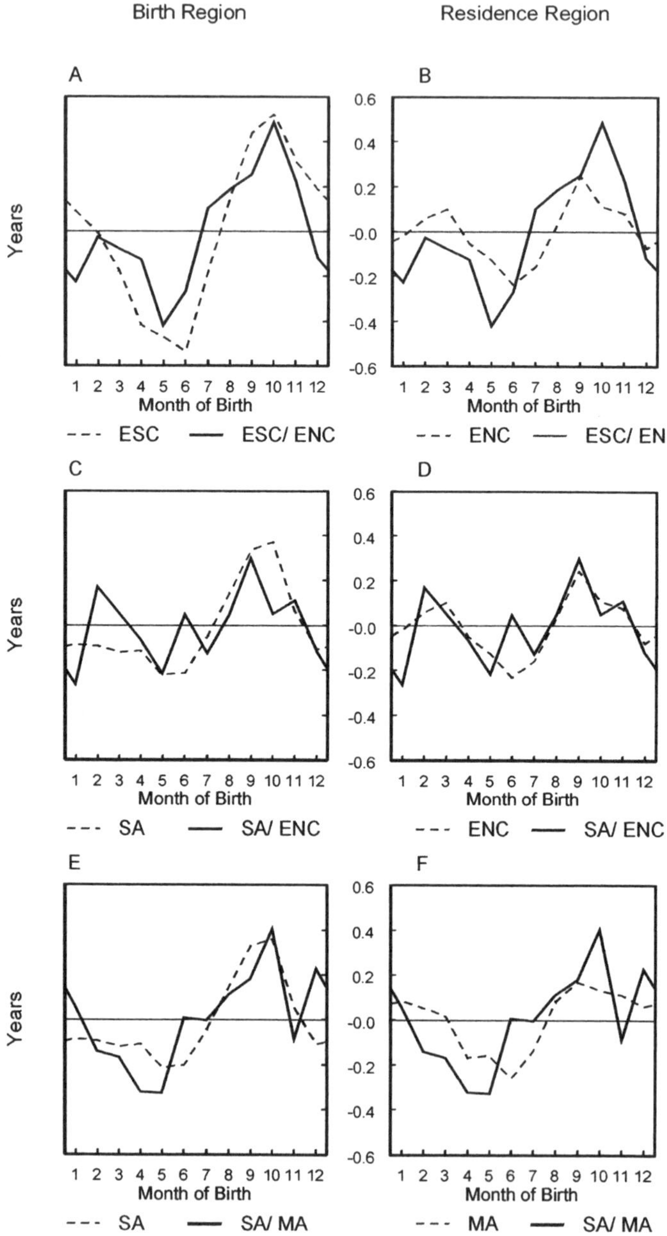

Figure 7.3. The month-of-birth pattern for white migrants from the South to the North: differences in mean age at death by month of birth from the average mean age at death at ages 50+. A,C,E) non-migrants' birth region (dashed lines), migrants (solid lines), B,D,F) non-migrants' residence region (dashed lines), migrants (solid lines).

migrants from *South Atlantic* to *East North Central* is significantly correlated with the pattern of the non-migrants of the residence region ($\rho=0.748$, $p=0.005$); it is not correlated with the pattern of the birth region ($\rho=0.413$, $p=0.183$). There is no difference in the correlation of the migrants from *South Atlantic* to *Middle Atlantic* with their birth region ($\rho=0.678$, $p=0.015$) and their residence region ($\rho=0.655$, $p=0.021$).

Black South-to-North Migration

A sizeable proportion of the South-to-North migration consists of African-Americans who moved north. Combining the three southern regions into the South and the four northern regions into the North, no significant differences are found between the month-of-birth patterns of the migrants, of those who remained in the South, and of those who were born in the North. The non-significant result is mainly caused by the large standard errors of the mean age at death of those born in the North. The correlation coefficients between the three groups show that the pattern of the migrants is more strongly correlated with that of the southern non-migrants ($\rho=0.804$, $p=0.002$) than with the pattern of those who were born in the North ($\rho=0.650$, $p=0.022$).

The evidence presented above suggests that both the month-of-birth

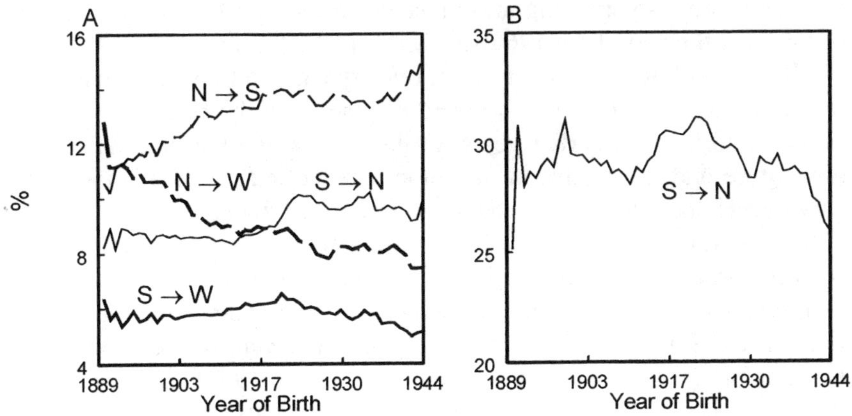

Figure 7.4. Proportion of migrants by year of birth and region of birth in the US death records 1989 to 1997 for whites (A) and for black South-to-North migrants (B).

patterns of migrants from Britain to Australia and of migrants within the US depends on their region of birth rather than on their region of residence. However, these results are not always consistent and in the reminder of the chapter we explore the possibility that unobserved social characteristics might be responsible for these inconsistencies.

In the North-to-South migration retirement migration has intensified over the years. This can be clearly seen in the proportion of migrants among the decedents of a given birth year (Fig. 7.4, solid line): 11 % of those who were born in 1880 were North-South migrants, as compared to almost 16% of those who were born in 1946. On average these migrants are more likely to be married, to have higher income and education levels and to be healthier than the non-migrant population (Biggar 1980, Speare & Mayer 1988). The above analysis accounts for the effects of family status and education and, thus, is able to compare the migrants with the average population of their birth region and their residence region.

South-to-North migration is labour-related migration. The death data clearly depict the different migration waves. Figure 7.4 shows the proportion of migrants among the decedents by year of birth. When one adds the average migration age (20–25 years) to the year of birth, then Figure 7.4 reflects the continuous South-North migration flow during the 1920s – a period of loss of agricultural jobs in the South – and during the Great Depression in the 1930s, which was a time of general impoverishment of the rural southern population. A migration upsurge occurred after World War II. The South-to-North migration peaked in the 1950's and a third wave occurred at the end of the 1960's.

Those who left were mainly farm owners who had lost their land, share tenants, and sharecroppers who were displaced by the boll weevil infestation of cotton in the earlier years. In the later years, the mechanisation of agriculture and the reduction of cotton acreage led to there being a large rural surplus population that had no other options than to migrate (Fligstein 1981). Although the large majority of the South-to-North migrants were born in rural areas they were by no means a homogenous group. On the contrary, large social differences existed among them. Social-class relations in the South, particularly in the cotton growing regions, were structured along the lines of land ownership (Fligstein 1981). This system distinguished between plantation owners and merchants on the one hand and farm labourers, tenants, and small farm owners on the other hand. Farm labourers occupied the lowest rank in this class system. These people were wage labourers and worked under the direct supervision of a farm owner. Tenants were divided into sharecroppers, share tenants, and cash tenants. They are distinguished by the amount of control imposed by the plantation owner or merchant. Sharecroppers were often almost in the position of the wage hand, as the landlord told the sharecropper what to plant and when to plant it. Share tenants were closer to being true renters – the landlord did not have the right to supervise their daily activity. Cash tenants had the greatest degree of independence.

For the analysis of the month-of-birth pattern it is important to recognize that farm labourers, sharecroppers, and to some extent also share tenants often did not have the means for subsistence farming and only planted cash crops. This implies that, for their food supply, they were fully dependent on their landlord or the merchant, which resulted in an extremely deficient diet well into the 1930s, known as the three m's: meat, meal (corn) and molasses (Levenstein 1993, 1998).

One can argue that the unobserved social differences in the rural South are the main reason for the inconsistencies in the South-North migration. This hypothesis is explored in the remainder of this chapter.

7.2.3 The Month-of-Birth Patterns of Farmers, Non-Farmers, and Farm Workers

Preston et al. (1998) showed that a rural place of birth and, in particular, a farming background positively influences the old-age survival of African-Americans. Since the US death certificates do not indicate whether one was born in a rural or urban area, indirect information is used. The usual occupation of the decedent offers alternative information. The occupation "farmer" usually implies a rural place of birth and in addition, the death records provide a distinction between farmers and farm workers.

Two competing hypotheses can be formulated concerning the month-of-birth pattern in urban and rural environments. The first hypothesis is based on the fact that infant mortality in rural areas was lower than in urban areas at the beginning of the 20th century (Preston & Hayens 1991). Infant mortality was mainly the result of infectious diseases, which spread more easily in urban areas due to crowded conditions in combination with the lack of sanitary infrastructure. Thus, if the month-of-birth pattern is primarily caused by infectious disease, then the swings in the pattern should be smaller in rural areas. On the other hand, nutrition was better in urban areas at the time (Dirks & Duran 2001) thanks to better means of preservation and transportation and because of the desperate situation of the sharecroppers and share tenants, particularly in the South. If nutrition plays a major role, then the differences should be larger among the rural population. The regression model that was fitted to the mean ages at death of farmers and non-farmers did not find any significant differences (results not shown). However, farmers tend to exhibit larger differences in the

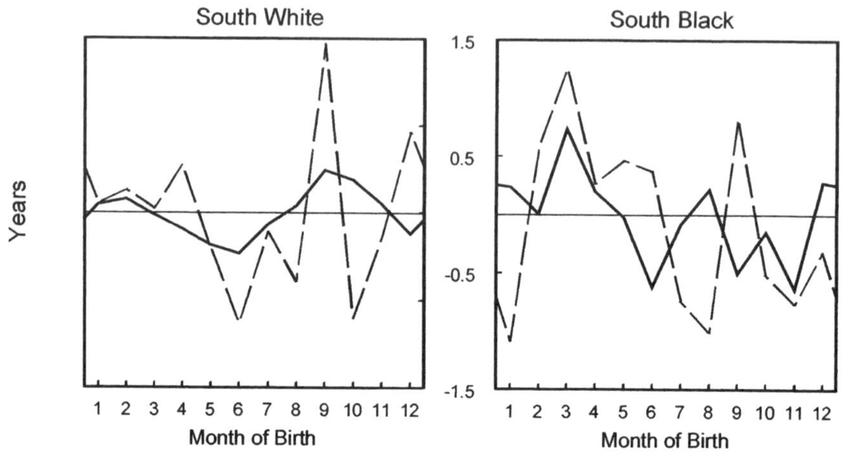

Figure 7.5. Age-adjusted differences in mean age at death of Southern farmer and farm workers by month of birth.

month-of-birth pattern than non-farmers both in the North and the South. No difference exists for blacks in the South.

The difference between farmer and farm worker may reflect the social stratification of the South better than education does. Since farm workers probably stem from landless rural families, their early-life environment must have been worse than that of farmers. The expectation therefore is that the peak-to-trough difference in the month-of-birth pattern of farm workers is larger than among farmers. In the North and the Mountain region the month-of-birth pattern does not differ significantly between the two occupational groups. In the South, however, the peak-to-trough difference for white farmer worker is significantly larger than for white farmer (p=0.037). There is no significant difference for black males (Fig. 7.5).

7.3 Conclusion

The month-of-birth pattern of migrants differs from the pattern of people born in their new residence region. British migrants to Australia and US internal migrants from the North to the South clearly show a distinctly different month-of-birth pattern than native-born Australians and US Southerners. However, the month-of-birth patterns of migrants also contain unexplained features such as the secondary summer peak in mean age at death among the British migrants. In addition the pattern of the US South to North migrants does not conform to the expectation at all.

At least two possible explanations exist for these phenomena. First, a large body of literature suggests that migrants are selected for better health. Since month of birth may affect not only mortality but also health, the tendency to migrate may depend on the month of birth, which would confound the results.

The second explanation is that migrants are socially selected and therefore do not represent their population of origin. The US South to North migration is a good example. Throughout this analysis education is generally used to account for the effects of social stratification. Education, however, is a poor measure of social class for the South in the first part of the 20th century. The primarily agricultural South was stratified not by education but by land ownership. The overall month-of-birth pattern of Southern migrants may thus largely depend on whether farmers (or sons and daughters of farmers) dominated the migration flow, or landless labourers who worked as farm workers, or migrants from urban areas.

This latter explanation can only be explored indirectly on the basis of the US death records because they do not contain information related to the social class of the parents of the deceased. One possibility to circumvent this lack of information is to use the information about usual occupation, which a subset of the US states reports in their death certificates. One occupation, which tends to be largely inherited, is farmer. The assumption is that farmers were mainly born into a farming background and that the usual occupation of the decedent, therefore, reflects the social status of the parents and therefore his/her own social status early in life. The death records distinguish between farm workers and farmers and one would expect that landless farm workers rather stem from landless rural parents than from a family that had once own land.

The comparison of the month-of-birth patterns of farmers and farm workers with the residual group non-farmers does not lead to conclusive results. The mean ages at death tend to exhibit a larger peak-to-trough difference among farmers than among non-farmers, multivariate analysis, however, shows that this difference is not significant. In the North, the difference in the month-of-birth pattern between farmers and the rest of the decedents is mainly due to social differences reflected in education. As soon the model corrects for educational differences, no further differences remain. In the South and the mountain region no significant differences in the month-of-birth pattern between farmers and non-farmers exist and education has no impact on this difference.

The most likely explanation for this finding is that a considerable proportion of the residual group "non-farmer" may have actually been born on farms or, more generally, in rural areas. Thus, existing differences cannot be identified based on the specification used in this study. This explanation might be particularly true for the South, where the boll weevil infestation, the Great Depression, and the transformation of Southern agriculture during the first part of the 20th century forced many small farm owners and tenants off their land.

In the agricultural South, significant differences in the month-of-birth pattern of white farmers and farm workers exist, as expected. White farm workers were particularly disadvantaged in the South in terms of their social position since the large majority of whites used to own their farm.

No significant differences in the month-of-birth pattern between farmer and farm worker exist for blacks. This can be explained by the fact that the large majority of black farmers were sharecroppers or share tenants with no freedom to decide what and when to plant. Like farm workers, they depended on their landlord for food and had no means for subsistence farming. In other words, among blacks there was not a lot of difference in the social status of farm workers and farmers.

Returning to the original question why the month-of-birth patterns of the South-to-North migration flows of whites does not follow the expectation it appears that unaccounted social differences may be largely responsible for the inconsistent results.

8 Does the Month-of-Birth Effect Exist in Cohorts Born Today?

8.1 Introduction

The fetal-origins hypothesis (Barker 1995) suggests that low birth weight and fast catch-up growth in the first months of life are associated with higher susceptibility to cardiovascular disease later in life. Central to this hypothesis is the reported inverse relationship between birth weight and systolic blood pressure later in life (for a review see Joseph & Kramer 1996). However, a recent study has cast doubt on whether there exists an association at all (Huxley et al. 2002).

The fetal-origins hypothesis uses birth weight as a surrogate measure of fetal nutrition, and malnutrition *in-utero* and/or exposure to infectious disease have been proposed as the causal mechanisms that link birth weight to later blood pressure (Barker 1989, Eriksson et al. 2000, Forsen et al. 2000, Eriksson et al. 2001). Other studies show that genetic factors may contribute to the relationship between blood pressure and birth weight (Christensen et al. 2001).

The previous chapters provide evidence that life span after age 50 depends on the month of birth and that seasonal differences in nutrition and the incidence of infectious diseases during the pregnancy of the mother or in the first months of life are the most likely causal factors. Nutrition and the incidence of infectious diseases vary seasonally. Thus, in the context of the fetal-origins hypothesis one would expect that seasonal fluctuations in nutrition and infectious-disease incidence should be reflected in seasonal fluctuations in birth weight. Taking this idea one step further, one would also expect to find a month-of-birth pattern in factors related to later-life health such as current weight, current height and current systolic blood pressure.

The fetal-origins hypothesis claims that nutrition and infectious diseases during the pregnancy of the mother are responsible for an increased susceptibility of the child to heart disease once it reaches adult ages. This claim, however, has been repeatedly challenged (Kuh & Davey-Smith 1993). Deprivation *in-utero* and during childhood may simply be the starting point of deprivation throughout life. The increased susceptibility later in life may simply be the result of the life-long accumulation of detrimental effects.

The season of birth does not have any life-course interpretation. Differences in mortality and health-related measures later in life by season of birth indicate that the fetal environment and the first months of life influence the susceptibility to chronic disease later in life independently of other life-course factors.

To test whether the seasonal changes in the early-life environment affect later-life health, there are two hypotheses. The first hypothesis is that seasonal fluctuations in nutrition and in the incidence of infectious diseases are reflected in a seasonal birth weight pattern and that this pattern is correlated with the season-of-birth pattern in systolic blood pressure, current weight, and current height in their young adult years. It is tested on the basis of the Minnesota twin family study. The second hypothesis is that the month-of-birth pattern in twins' birth weight is positively correlated with the differences in mean age at death by month of birth for decedents aged 50+ who were born in Minnesota. If this is the case, then it could suggest that the differences in life span by month of birth also exist in contemporary cohorts.

Twins are exposed to harsher conditions *in-utero* than singletons because they have to share resources. It has been argued (Phillips 1993) that even small differences in the birth weight of twins could reflect important differences in intrauterine conditions because twins already have considerably lower birth weights than singletons. Since the interest is primarily in the effect of month of birth in contemporary populations, these seasonal effects may be more readily detected among twins than among singletons because they should be larger among twins.

Evidence concerning the relationship between month of birth and birth weight, length at birth, current adult and adolescent height and weight exist for different regions of the world and for different time periods (Table 8.5 at the end of this chapter). This evidence is reviewed briefly below. Only few of these studies have information on all these measures, however, and many are based on historical populations or populations from developing countries. Only one study concerning the effect of month of birth on blood pressure exists. The study is based on Spanish men aged 45-64 and finds a significant difference of 11.1 mmHG between the minimum in systolic blood pressure among the spring-born and the maximum among the fall-born (Banegas et al. 2000).

8.1.1 Birth Weight

Plenty of evidence exists concerning seasonality in birth weight. The studies range from contemporary populations in tropical and sub-tropical regions among whom malnutrition is endemic to contemporary populations in Australia, Japan, Ontario, and Denmark and to historic births between the end of the 19th century and the middle of the 20th century in Germany and Austria. Equally important are studies on weight gain in developing countries which demonstrate that the seasons of the year significantly influence both birth weight and the rate of weight gain during infancy and childhood.

In tropical and sub-tropical regions, hot and rainy summers are usually associated with low birth weight (Pollit & Arthur 1989, Roberts et al. 1982, Bantje & Niemeyer 1984, Kinabo 1993, Wendl-Richter 1997), and slow weight gain in infants and children (Brown et al. 1982, Chen et al. 1979) and weight loss in women (Chen et al. 1979, Cole 1993).

In Bangladesh (Brown et al. 1982) the percentage of expected monthly gain in infant weight or length varied by a factor of three or four during different months of the year. The most acute growth faltering was observed during the hot and rainy period in August and September. On the other hand, both the velocity of infant growth and women's weight increase in cold dry winters, from about February to June.

A longitudinal study of infants born in rural Taiwan showed that birth weights were lowest during the hot and wet months of June, July, and August (Adair & Pollit 1985, 1983). The authors did not find a statistically significant effect in weight gain during the first trimester by month of birth. However, the summer cohorts, despite their lower birth weights, tended to surpass the weight of infants born in other months for a greater part of the remainder of the first year. This is mainly due to the higher weight gains in autumn and winter. By the age of 12 months, there were no significant differences in weight between any of the four cohorts.

Similar results are obtained in a study of infants born in Shanghai (Xu et al. 2001). In particular, weight velocity was higher in autumn and winter than in summer in all age groups. The study fails, however, to report whether differences in birth weight exist.

Seasonality in birth weights has also been observed in contemporary populations of developed countries. A study of indigenous Australians demonstrates that significantly more babies with a very low birth weight (<1500g) were born during the wet season (Rousham & Gracey 1998) than during the dry season, while there was no significant seasonal difference in births of 1500-2499g. The authors conclude: "since very low birth weight babies have a strong likelihood of being pre-term the findings suggest that seasonality of birth weight may be due to the increase in pre-term births rather than an increase in rates of intra-uterine growth retardation."

A Japanese study of all births occurring between 1974 and 1983 found similar significant differences in mean birth weight and mean birth weight at 40 weeks of gestation (Matsuda et al. 1995). The pattern is bimodal, with peaks in spring and autumn and troughs in winter and summer. These results indicate that both pre-term births and the intrauterine growth rate contribute to the seasonality in birth weight.

A Danish study of gestational age, length, and weight at birth of all infants born between 1973 and 1994 found seasonal variation in mean length and weight at birth. It fails, however, to report the exact pattern of the differences in birth weight (Wohlfahrt et al. 1998).

A comparison of seasonal variations in birth weights between rural Zaire and Ontario, Canada, (Fallis & Hilditch 1989) did not find a statistically significant seasonal variation in birth weight in Ontario. This is also true for 1,750 men and women born in Hertforshire between 1920 and 1930, where the authors failed to find any significant seasonality in birth weight. However, the number of subjects included in the study might be too small to find significant results.

Seasonality in birth weight has been widely studied in Germany on the basis of historical data. In a study of 101,444 births in Berlin and Leipzig (Otto & Noack 1957) for the years 1936-1953 (Berlin) and 1937-1953 (Leipzig) the authors find that infants born in autumn and early winter had higher birth weights that those born in the first half of the year. The difference is about 35g. In the same study they report differences in birth weight on the basis of data from Marburg, Vienna and Odessa over the years 1920-1922. In this dataset the difference between the peak in mean birth weight in the autumn and the trough in February is more than 60g.

Based on 26,515 births in Hamburg in the time period 1912 to 1922, Hellmuth & Wnorowski (1923) concluded that the autumn-born have higher birth weights, although the differences are not significant. A study of weight at birth in Vienna between 1865 and 1930 (Ward 1987) shows a difference of 58g between infants born in winter and spring (trough) and those born in summer and autumn (peak).

8.1.2 Current Weight

Fewer studies have been conducted on the relationship between current weight and month of birth. An early study was carried out by Hillman and Conway (1972) involving 9,103 patients who attended a public-health nutrition clinic in the United States during the 5-year period 1966 to 1970.

They found a significant deficit of the birth months October and November ($p<=0.01$) among the 5,627 overweight patients and a single peak in March. This pattern was most pronounced in patients aged 20 to 30. Although the overall patterns for underweight and normal-weight patients did not vary significantly, they found a relative deficit of birth dates during June and July ($p<0.05$). The authors suggest that the relationship between month of birth and being overweight changes with age, "an impression consonant with the concept that the pathogenesis of later acquired (adult) obesity commonly differs from the type that is manifested in childhood".

Hillman and Conway observed that, under the age of 20, people born in spring and early summer seem less apt to be overweight. Above the age of 20, this is true for those born in the second part of the year. They mention, however, that their results are only partly consistent with other studies about season of birth and body form (Chenoworth & Canning 1941, Mills 1941, Fitt 1955). The study by Fitt (1955) is based on data from New Zealand. He reports the mean height and weight of adult men in the New Zealand army by month of birth. Taking their weight for height ratio, one finds a bi-modal pattern. Those born between February and July and in October-November are slimmer than those born in August, September, and January.

In a study of 2,500 young women, those born in September and October show the highest weights (Hillman et al. 1970). A recent study about birth weight, climate at birth, and the risk of obesity later in life (Phillips & Young 2000) studied 1,750 men and women born in Hertfordshire between 1920 and 1930. Average age at the time of the interview was 64.9 years. Among men, the prevalence for obesity varied with month of birth and was greater among those born in January to June than among those born in July to December. Among women there was no specific trend.

Several studies on the relationship between current weight and birth weight have been performed which do not consider the month of birth. A study of 14-year old Polish adolescents (Koziel & Jankowska 2002) shows that boys who were small for their gestational age at birth (SGA) had a lower body mass index at age 14, while SGA girls had accumulated more centralized fat. In a study of 3,447 women born in Helsinki who developed coronary heart disease (Forsen et al. 1999) the mean heights and weights at ages 7 to 15 of those who developed coronary heart disease were below those of all other women. They had, however, experienced catch-up growth in childhood since at birth the difference in height and weight was larger than in adolescence. A similar study for males (Forsen et al. 1997, Erikson et al. 1999) finds the highest death rates from coronary heart disease later in life among boys who were thin at birth and who had caught up in weight by the age of seven.

In conclusion, the studies on current weight and weight gain indicate that a relationship with the month of birth does indeed exist. However, this relationship is complicated by the fact that it seems to differ with age and for the two sexes. The latter fact might be explained by differences in the tempo of growth during adolescence between boys and girls.

8.1.3 Length at Birth and Height Gain During the First Year of Life

Many studies about seasonal differences in height by month of birth look at height gain during the first years of life or at length at birth. For example, Xu et al. (2001) found that, in Shanghai, month of birth has some association with attained size but that this is reduced during the first two years of life. A study of infants born in Gambia arrived at similar results (Cole 1993). All the studies confirm the general knowledge reflected in the German saying that " *Mairegen laesst die Kinder wachsen* ("Rain in May lets children grow)", i.e. that height gain differs by the seasons of the year. This implies that infants and children of the same age grow at different rates depending on their month of birth.

There is one contemporary study about differences in length at birth (Wohlfahrt et al. 1998). It is a population-based study of 1,166,206 children born in Denmark between 1973 and 1994. Children born in April were 2.2 mm longer than those born in December. In their study of births in Leipzig and Berlin Otto and Noack (1957) report both the seasonal distribution in birth weight and length at birth. They find that length at birth peaks in March, April, and October and reaches a trough in May and June. In a second article Otto and Glaas (1959) use the same data from Leipzig and calculate mean gestational age, mean birth weight, and mean length at birth by month of birth. On the basis of these three measures it appears that, corrected for their gestational age, the autumn-born infants are tall and heavy while the spring-born infants are tall and thin.

8.1.4 Current Height

As early as 1941 Fitt analysed the relationship between month of birth and current height among children and adolescents (Fitt 1941) and came to not always consistent results with other contemporary studies (Mills 1941, Chenoweth & Canning 1941). Fitt attributes this to the different ages of the adolescents and to the fact that they were still growing at different rates. In his 1955 article he presents results for the heights of 21,342 adult men from a large sample of men drafted into the New Zealand army in World

War II. Arranging the height measures by month of birth, he finds a cyclical pattern. Those born in June are the shortest, and the tallest are born between September and February. The peak-to-trough difference in height is 0.31 inch=0.8 cm. Turning to the Northern Hemisphere, Weber et al. (1998) report a sinusoidal cycle for the heights of the entire male population aged 18 for ten one-year birth cohorts. The Austrian Federal Army measured these cohorts throughout the whole year in five conscription centres. They find that the tallest are born in March and the shortest in September. The peak-to-trough difference is 0.6 cm. These results are consistent with a study by Breitinger (1966), who found a similar pattern for those born in January to June and July to December. A Spanish study (Banegas et al. 2001) of 2021 men and women found that males born in summer were 1.7 cm taller than those born in winter. This difference was most marked for those with manual occupation. No differences existed for females.

The past studies about month of birth and height lead to three conclusions: First, the spring-born seem to be the longest infants at birth. There is evidence, however, that they are long and thin while the autumn-born are long and heavy. Second, due to the seasonality in the velocity of growth, infants and children of the same age have very different heights according to their month of birth, which complicates the analysis. Third, at adult ages the spring-born are the tallest in the Northern Hemisphere while the pattern is shifted by a quarter of a year in the Southern Hemisphere.

8.2 Data and Method

This study is based on adolescent twin participants of the same sex from the Minnesota Twin Family Study. Two cohorts are assessed: the first consists of twins born between 1977 and 1984 who were approximately 11 years old (age range 11 to 12 years) at the time of the interview, the second consists of twins born between 1972 and 1979 aged approximately 17 (age range 17 to 18 years). A description of the Minnesota Twin Family Study and how the measurements were obtained can be found in Christensen, Støvring and McGue (2001).

The sample used in this study consists of 1,353 twin-pairs (660 male, 693 female). Twin-pairs with missing measurements and outliers are excluded. Outliers of height, current weight, current systolic blood pressure and birth weight are identified on the basis of the inter quartile range (IQR) and the first (q1) and third (q3) quartile. Those measures are classiefied as extreme outliers that either exceeding q1-3*IQR or q3+3*IQR. Table 8.1

Table 8.1. Number of twin pairs used in the cosinor analysis.

	Current weight	Current height	Current systolic BP	Birth weight
Ages 11-12				
Males	347	359	346	363
Females	340	350	348	347
Ages 17-18				
Males	268	273	274	273
Females	299	308	305	305
Both Ages				
Males	615	632	620	636
Females	639	658	653	652

contains the number of twin pairs that are used in the different calculations.

A total of 340,328 US death records for the years 1989 to 1998 were used to calculate mean age at death by month of birth for people born in Minnesota. The data are publicly available from the Multiple Cause of Death Data from the National Center for Health Statistics (NCHS), to which the information about month of birth was added.

In the twin data the mean of each outcome measure is calculated by week of birth. Cosinor analysis is used to test whether the week-of-birth pattern is statistically significant and whether the fluctuations are seasonal. For each of the four measures birth weight, current weight, current height, and current systolic blood pressure the following model is estimated.

$$y = \alpha_0 + \alpha_1 \sin(t) + \alpha_2 \cos(t) + u$$

$$t = \begin{cases} \text{week of birth} / p * 2\pi & \text{if one peak} \\ \text{week of birth} / \dfrac{p}{2} * 2\pi & \text{if two peaks} \end{cases} \qquad [8.1]$$

In Equ. 8.1 y denotes one of the five outcome measures, α_1 and α_2 are regression parameters, α_0 is the general mean and t is the week of birth defined in the known period of 2π; p is the number of time units in period 2π. The error term u follows a normal distribution with mean zero and variance σ^2.

For each measure three models are estimated: the first model tests whether an unimodal sinusoidal function fits the data and defines the period 2π in terms of 52 weeks (i.e. p=52); the second model tests for a bimodal sinusoidal function and assumes that the period is defined over p=26 weeks. The third model tests whether the sum of a unimodal and a bimodal sinusoidal function fits the data best. Equ. 8.1a allows for modeling of seasonal fluctuations that depart from pure sinusoidal functions.

$$y = \alpha_0 + \alpha_1 \sin(t) + \alpha_2 \cos(t) + \alpha_3 \sin(t_1) + \alpha_4 \cos(t_1) + u$$

$$t = week\ of\ birth / p * 2\pi \qquad [8.1a]$$

$$t_1 = week\ of\ birth / \frac{p}{2} * 2\pi$$

Equation 8.1 is the linearized version of an equivalent formulation with slightly different parameterization:

$$y = \beta_0 + \beta_1 \cos(t - \beta_2) + u \qquad [8.2]$$

It has been shown (Bingham et al. 1982) that

$$\beta_1 = \sqrt{\alpha_1^2 + \alpha_2^2}$$

$$\beta_2 = \arctan\left(\frac{\alpha_q}{\alpha_2}\right) + k \qquad [8.2a]$$

$$\beta_o = \alpha_0$$

In cosinor analysis β_0 is named mesor and estimates the general mean, β_1 is the amplitude and β_2 is called acrophase, the point were the sinusoidal function reaches its maximum/minimum.

US death data contain information about the date of birth on the level of month of birth but not by week of birth. Therefore, all calculations are also performed on a monthly basis with p=12.

The final model is selected on the basis of the Akaike information criteria max (2LL-2p), where LL denotes the log likelihood and p the number of parameters in the model. Since twin data are correlated *Generalized Estimating Equations* (GEEs) based on Liang and Zeger are used to estimate the models. The GEE method is implemented through the "REPEATED" statement of the "GENMOD" procedure in SAS.

Table 8.2. Parameter estimates and p-values of the cosinor analysis of birth weight, current weight, current height and current systolic blood pressure.

		Birth weight	Current weight	Current height	Current systolic blood pressure
Boys					
11-12	Periodicity	26	52	52	26
	α_1 *(sin)*	-.06	**4.27**	**2.33**	1.03
	p-value	(.13)	(.01)	(.00)	(.18)
	α_2 *(cos)*	-.07	-1.66	**-1.05**	-1.18
	p-value	(.07)	(.20)	(.03)	(.07)
17-18	Periodicity	26	52	26	52
	α_1*(sin)*	-.05	-.91	**1.23**	-1.34
	p-value	(.25)	(.61)	(.05)	(.17)
	α_2 *(cos)*	**-.09**	-2.71	-.17	.86
	p-value	(.04)	(.15)	(.76)	(.35)
Girls					
11-12	Periodicity	52	52+26	52	52+26
	α_1 *(sin)*	-.03	**3.70**	**2.64**	1.43
	p-value	(.47)	(.03)	(.00)	(.17)
	α_2 *(cos)*	-.04	**-3.40**	**-1.29**	**3.22**
	p-value	(.26)	(.03)	(.01)	(.00)
	α_3 *(sin)*		**3.59**		-1.65
	p-value		(.03)		(.11)
	α_4 *(cos)*		-0.79		**-2.71**
	p-value		(.60)		(.01)
17-18	Periodicity	26	26	52	52
	α_1 *(sin)*	-.05	0.72	-.17	**1.0**
	p-value	(.22)	(.64)	(.73)	(.05)
	*α_1 *(sin)*	-0.05*			
	**p-value*	(.16)*			
	α_2 *(cos)*	-.04	**-3.49**	-.70	1.63
	p-value	(.40)	(.03)	(.09)	(.11)
	*α_2 *(cos)*	**-0.08** *			
	**p-value*	(.05)*			

1. Significant values are set in bold
2. Periodicity: 26 weeks indicate one bimodal fluctuation; 52 weeks one unimodal fluctuation; 52 + 26 weeks the sum of one bimodal and one unimodal fluctuation
3. The best fitting model is chosen based on the Akaike information criteria
4. * refers to the parameter estimates based on the definition of outliers: q1-1.5*IQR or q3+1.5*IQR

Table 8.3. Predicted peaks and troughs in birth weight, current weight, current height and current systolic blood pressure based on Equ. 8.1 and 8.1a of twins aged 11-12 and 17-18.

		Weeks Males/Females			
		Birth Weight kg	Current Weight kg	Current height cm	Current systolic blood pressure Hg
Ages 11-12	Periodicity	26/..	52/52	52/52	26/52+26
	Maximum1	16/..	16/10	16/17	10/13
	Maximum2	42/..	../30	../..	36/44
	Minimum1	3/..	42/21	42/43	49/2
	Minimum2	29/..	../46	../..	23/29
	Max diff.	0.19/..	4.16/6.31	5.10/5.87	3.14/10.19
Ages 17-18	Periodicity	26/26	../26	26/..	../52
	Maximum1	15/17	../12	7/..	../7
	Maximum2	41/43	../38	33/..	../..
	Minimum1	2/4	../25	20/..	../33
	Minimum2	28/30	../51	46/..	../..
	Max diff.	0.20/0.12	../3.56	2.48/..	../5.06

8.3 Results

Twins Ages 11-12

Significant month-of-birth patterns exist for all four measures (Table 8.2) and they are generally similar for males and females. Current weight and current height each follow a unimodal pattern with a peak in spring and a trough in fall (Figures 8.1 and 8.2). They are highly correlated and it is obvious that the seasonal fluctuations in current weight are caused by the seasonal fluctuations in height. The differences between the peaks and troughs are considerable. Among boys the spring-born are 5.10 cm taller and 4.16 kg heavier than the fall-born, the difference among girls is 5.87 cm and 6.13 kg (Table 8.3). For males current systolic blood pressure follows a bimodal function with peaks in spring and fall and troughs in summer and winter and a difference of 3.14 Hg (Figure 8.3). For females there

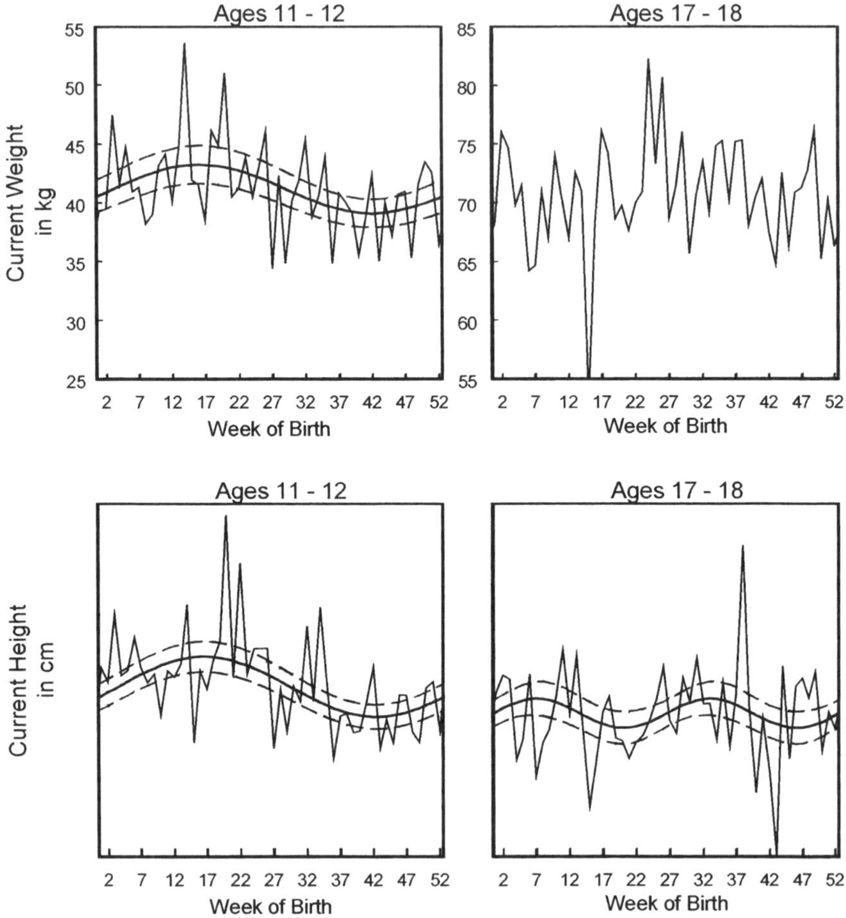

Figure 8.1. Mean current weight and current height by week of birth and sinusoidal functions incl. 95% confidence bands estimated by Equ. 8.1 and 8.1a, males, ages 11–12 and 17–18.

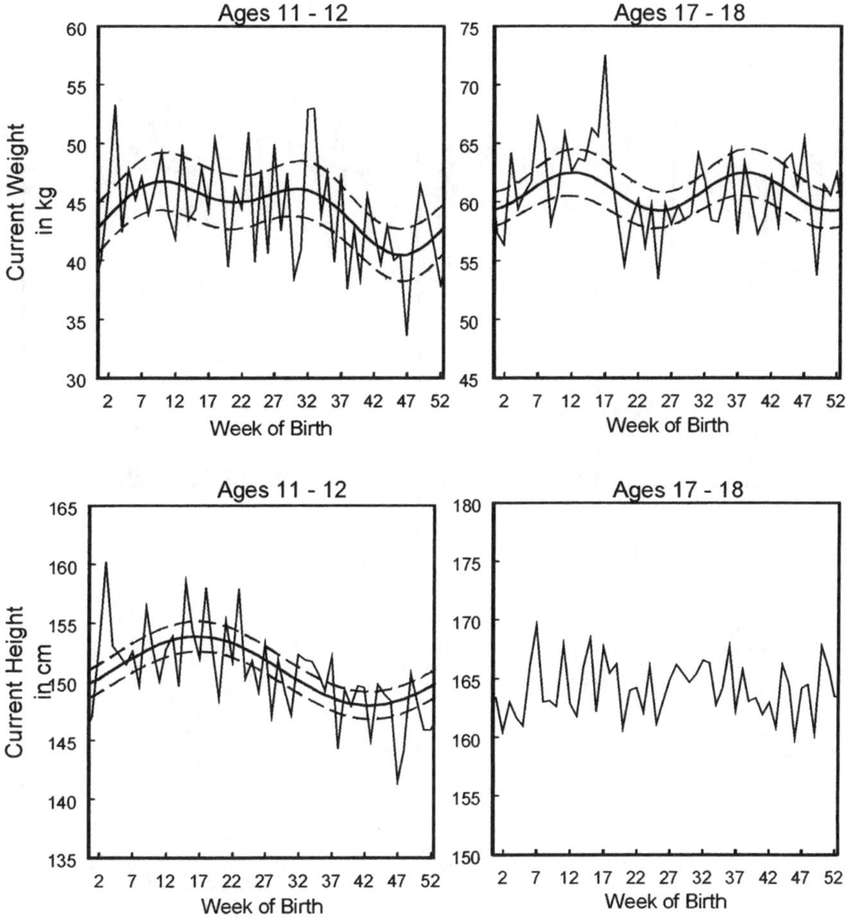

Figure 8.2. Mean current weight and current height by week of birth and sinusoidal functions incl. 95% confidence bands estimated by Equ. 8.1 and 8.1a, females, ages 11–12 and 17–18.

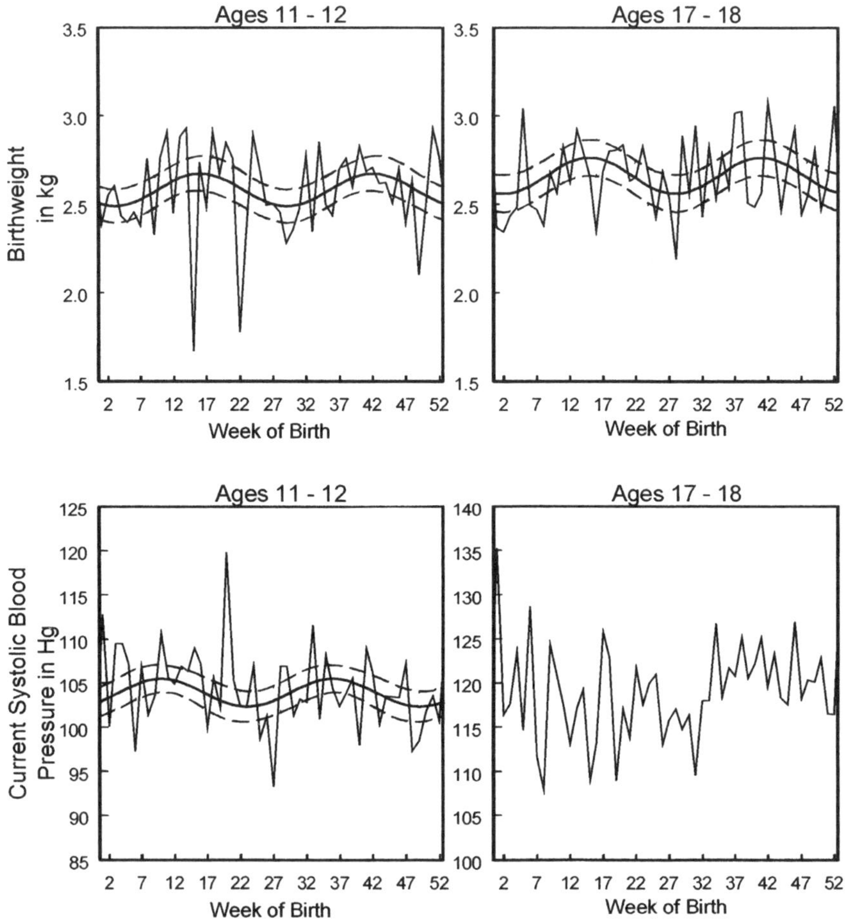

Figure 8.3. Mean birth weight and current systolic blood pressure by week of birth and sinusoidal functions incl. 95% confidence bands estimated by Equ. 8.1 and 8.1a, males, ages 11–12 and 17–18.

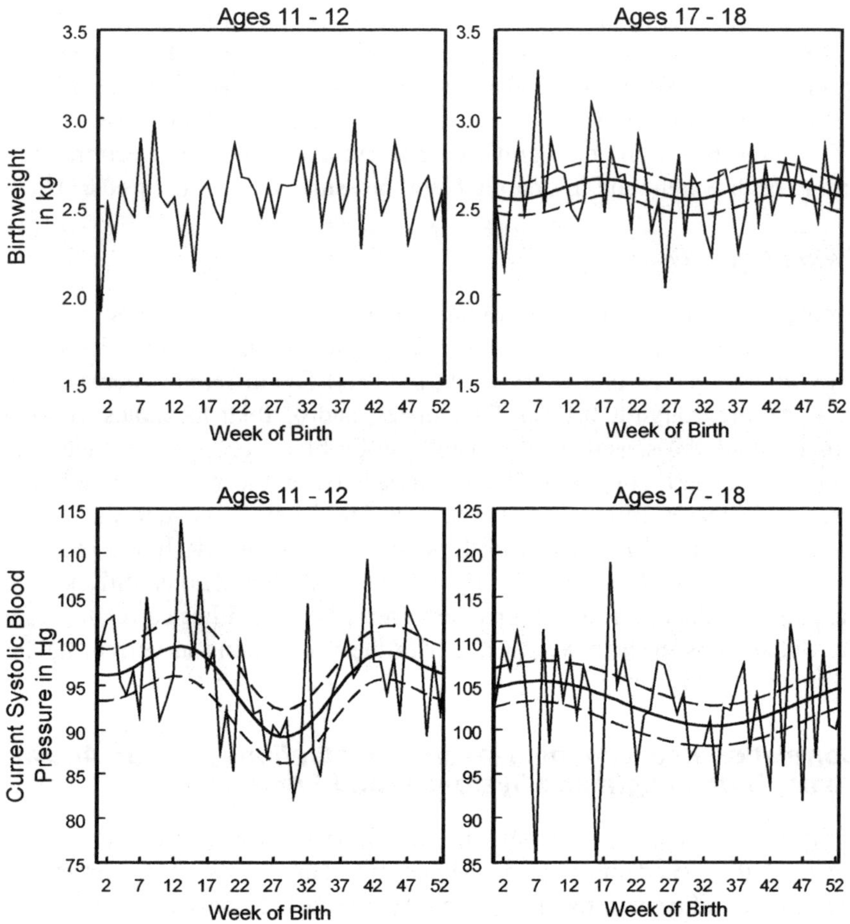

Figure 8.4. Mean birth weight and current systolic blood pressure by week of birth and sinusoidal functions incl. 95% confidence bands estimated by Equ. 8.1 and 8.1a, females, ages 11–12 and 17–18.

is a difference of 10.19 Hg between the trough among the summer-born and the peak among the winter-born (Figure 8.4). A striking seasonal pattern exists for male birth weight (Figure 8.3). It is bi-modal with peaks in spring and fall and troughs in summer and winter and a difference of 0.19 kg. No seasonal pattern exists in the birth weight of female twins. For males there is a high and significant positive correlation between the pattern in birth weight and in systolic blood pressure (PCC: 0.63, $p<0.03$).

Twins Ages 17-18

Among the older twins the seasonal pattern in birth weight is significant for both sexes and similar to the pattern of the younger male twins (PCC between ages 11-13 and 17-19: 0.99, $p=0.01$). The difference between the peak and the trough is 0.12 kg for females, and 0.20 kg for males. Among female twins the seasonal pattern in systolic blood pressure remains significant, although compared to the younger twins the peak-to-trough difference decreases from 10.19 Hg to 5.06 Hg. There is no correlation between the seasonal pattern in birth weight and in systolic blood pressure for the older female twins. Significant seasonal fluctuations still exist in the height of the older male twins, however they are neither correlated with the fluctuations in birth weight nor with the fluctuations in current height of the younger twins.

Correlation Between the Seasonal Fluctuations in Mean Age at Death, Birth Weight and Systolic Blood Pressure

Significant seasonal fluctuations by month of birth exist in mean age at death of the decedents born in Minnesota. This is also true for cause-specific mortality with the exception of the group of non-natural causes of death among males (Table 8.4.).

If one calculates the correlation between the 12 monthly values predicted by the sinusoidal patterns then a positive correlation exists between the sinusoidal fluctuations by month of birth in birth weight and mean age at death. The Pearson correlation coefficient is generally about 0.42 for both sexes, however, statistically it is only of borderline significance. Between systolic blood pressure and mean age at death the correlation is strong and significant among younger female twins (PCC=0.72, $p=0.009$).

Among males the sinusoidal fluctuations in total mortality, cancer, heart disease and other natural causes of death consist of a unimodal part defined over 12 months and a bimodal part defined over 6 months. The acrophases of the bimodal parts are highly consistent for total mortality, birth weight and systolic blood pressure: the values range from 0.95 for total

Table 8.4. Parameter estimates and p-values of the cosinor models of the effect of month of birth on total mortality, cause-specific mortality, birth weight and current systolic blood pressure.

Males	Minnesota death records					Minnesota twin study			
	Total	Cancer	Heart disease	Other natural	Non-natural	Birth weight ages 11-12	Birth weight ages 17-18	Systolic BP ages 11-12	Systolic BP ages 17-18
Periodicity	*12+6*	12+6	12+6	12+6	6	6	6	6	
α_1 (sin)	-.04	.01	.01	***.12***	n.s.	n.s.	n.s.	n.s.	n.s.
95% LCI	(-.11–	(-.14–	(-.09–	*(-.26–*					
95% UCI	.04)	.12)	.11)	*.02)*					
α_2 (cos)	**.19**	**.13**	**.18**	**.22**	n.s.	n.s.	n.s.	n.s.	n.s.
95% LCI	(.12–	(.05–	(.07–	(.09–					
95% UCI	.26)	.31)	.28)	.37)					
Acrophase β_2	0.79	1.08	1.06	1.50					
Amplitude β_1	0.39	0.26	0.36	0.50					
α_3 (sin)	**-.07**	.06	***-.10***	**-.15**	n.s.	**-.09**	**-.07**	.01	n.s.
95% LCI	(-.15–	(-.07–	*(-.21–*	(-.29–		(-.17–	(-.15–	(-1.07–	
95% UCI	-.00)	.19)	*.01)*	-.01)		-.01)	-.00)	-1.10)	
α_4 (cos)	-.05	**-.12**	-.02	-.05	n.s.	-.06	-.06	-1.44	n.s.
95% LCI	(-.12–	(-.25–	(-.13–	(-.18–		(-.12–	(-.14–	(-2.42–	
95% UCI	0.02)	.01)	.09)	.09)		.15)	0.02)	-.46)	
Acrophase β_2	0.95	0.54	1.37	1.25		0.98	0.86	0.99	
Amplitude β_1	0.17	0.27	0.20	0.32		0.22	0.18	2.95	

Table 8.4. (continued)

Females	Minnesota death records					Minnesota twin study			
	Total	Cancer	Heart disease	Other natural	Non-natural	Birth weight ages 11-12	Birth weight ages 17-18	Systolic BP ages 11-12	Systolic BP ages 17-18
Periodicity	12+6	12+6	12+6	12	6		6	12+6	12
α_1 (sin)	-.05	.01	.02	**-.14**	n.s.	n.s.	n.s.	**2.08**	**1.90**
95% LCI	(-.12–	(-.14–	(-.12–	(-.26--				(.17–	(.12–
95% UCI	.02)	.16)	.07)	.01)				3.99)	3.67)
α_2 (cos)	**.21**	**.24**	**.19**	*0.12*	n.s.	n.s.	n.s.	**3.04**	1.85
95% LCI	(.13–	(.08–	(.09–	*(-.01–*				(1.09–	(-.01–
95% UCI	.28)	.39)	.28)	*.24)*				4.99)	3.72)
Acrophase β_2	*.77*	*1.04*	*1.10*	*.14*				*0.60*	*0.80*
Amplitude β_1	*.43*	*.48*	*.38*	*.37*				*7.37*	*5.31*
α_3 (sin)	*-.06*	-.05	-.05	n.s.	-.28	n.s.	**-.07**	**-2.66**	n.s.
95% LCI	*(-.13–*	(-.20–	(-.15–		(-.83–		(-.13–	(-4.64–	
95% UCI	*.01)*	.09)	.04)		.26)		.00)	-.68)	
α_4 (cos)	**-.14**	**-.03**	**-.11**	n.s.	**-.56**	n.s.	-.03	-1.51	n.s.
95% LCI	(-.20–	(-.39–	(-.20–		(-1.11–		(-.11–	(-3.36–	
95% UCI	-.06)	-.09)	-.01)		-.01)		0.05)	.33)	
Acrophase β_2	*.41*	*1.03*	*.42*		*.46*		*1.16*	*1.05*	
Amplitude β_1	*.31*	*.12*	*.24*		*1.25*		*0.15*	*6.12*	

mortality, over 0.98 for birth weight of age group 17-18 (0.86 for age group 11-12) to 0.99 for systolic blood pressure at age 11-12 (Table 8.4).

For females the acrophases of the unimodal parts of the sinusoidal functions are 0.77 for total mortality and 0.80 for systolic blood pressure at age 17-18. The acrophases of the bimodal parts of the functions are consistent for cancer (1.03), birth weight of age group 17-18 (1.16) and systolic blood pressure at age 11-12 (1.05).

8.4 Conclusion

In the Minnesota twin data significant and consistent season-of-birth patterns exist in the four measures birth weight, current height, current weight and current systolic blood pressure. They are all consistent with patterns reported in earlier studies.

The season-of-birth patterns for measures related to current age are generally age dependent since they tend to become smaller or even to disappear with age. The large season-of-birth differences in current height and current weight that exist among the younger twins are generally absent at ages 17-18. There is one exception because current height among the older male twins still shows a seasonal month-of-birth pattern.

Recent studies suggest that at adult ages the spring-born men are taller. There is no information about females because the analyses are based on military conscript data. The differences found in these studies, however, are usually a few millimeters and therefore much smaller than those among the twins. Earlier studies have already suggested that the month-of-birth pattern depends on age. Inconsistencies in the results of studies of adolescents by Fitt, Mills, and Chenoweth and Canning were attributed to the different ages of the adolescents and that they were still growing at different rates. However, there is no study that particularly looked at the changes with age. The results from the twins suggest that the interaction between the seasonality in growth and the differences in the rate of growth with age is particularly strong at childhood while growth during puberty seems to have an equalizing effect.

A strong and consistent season-of-birth pattern with two troughs (summer and winter) and two peaks (spring and fall) exists for birth weight with the exception of the younger female twins. An important question is whether this pattern is caused by differences in gestational age and therefore by an increased risk of pre-term births, or whether it is caused by growth-retardation *in-utero*.

There is a general consensus that seasonal differences in birth weight in developing countries are due largely to intrauterine growth retardation. In studies on birth weight in developing countries, three main seasonal factors of intrauterine growth retardation are discussed. First, energy intake during the cold and dry period is about a third greater than during the pre harvest monsoon due to the availability of minor food crops (Brown et al. 1985). Second, rainfall and high temperatures influence disease prevalence during the hot and rainy period in that they have an effect on the incidence of infectious diseases such as gastrointestinal infections (Chen et al. 1979, Datta 1978, Trowbridge & Newton 1979), tuberculosis, measles, and whooping cough (Gordon 1965, Chambers 1981). Third, most agricultural activities are concentrated in the wet season, and the amount of work that women invest in the field increases by a factor of two to three (Brin 1984). The increase in work affects not only energy expenditure – it also reduces the time for social activities, including child rearing. There is an inverse relationship between the amount of time invested in farm work and time invested in activities in the household (Roberts et al. 1982).

There is no consensus yet as to whether the seasonal pattern in the birth weights of historic and contemporary western populations (including Japan) is due to differences in gestational age or to intrauterine growth retardation. The 1993 study by Matsuda et al. shows for Japan that, although the seasonal pattern is dampened, it nonetheless persists – even when corrected for gestational age.

Gestational age is not reported in the Minnesota twin family study, so it is not possible to control for gestational age. There is, however, a study of over 400,000 white singleton live births and stillbirths in Minnesota in the years 1967-1973 by Keller and Nugent (1983) which reports the seasonal pattern of preterm births (29-37 weeks gestational age). Like birth weight, it reveals two troughs and two peaks. The risk of a preterm-birth is particularly high in December and February. It reaches its minimum in March and another peak in July and August. The risk is higher in the autumn than in the spring. A highly significant correlation between the month-of-birth pattern in mean birth weight and the monthly probability of a preterm birth (ρ= -0.73, p=0.007) exists among the older female twins, whereas there is none among the younger female twins (ρ=0.091, p=0.78). For males a non-significant negative correlation exists for the older age group (ρ=-0.38, p=0.22), and there is a negative correlation of borderline significance for the younger age group (ρ=-0.54, p=0.07). The correlation coefficients thus suggest that the monthly differences in the birth weights of twins are partly the results of differences in gestational age. Since the correlation is modest, particularly among males, it also suggests that part of the differences are caused by growth retardation *in-utero*.

The comparison of the seasonal pattern in the risk of a pre-term birth reported in the study of Keller and Nugent and in birth weight reported by this study raises doubts about the validity of the results for the younger female twins. They are the only group without any significant seasonal differences in birth weight. It is highly unlikely that the seasonal differences in gestational age related to prematurity would not be reflected in seasonal differences in birth weight. The lack of any seasonality in the birth weight of the younger females twins is therefore most probably due to problems with the data.

A recent review (Huxley et al. 2002) questions the validity of the relationship between low birth weight and increased systolic blood pressure levels at adult ages. The authors show that the magnitude of the relationship depends on the size of the study population: the larger the population, the smaller is the correlation between birth weight and blood pressure. Furthermore, they point out that most of the studies are adjusted for current weight, which might produce a spurious inverse relationship even if current blood pressure and weight are uncorrelated (Luca et al. 1999, Kramer 1987).

In our study we find a significant positive correlation between the seasonal fluctuations in systolic blood pressure and birth weight for male twins aged 11-12. The correlation, however, is positive because the spring- and the winter-born have lower birth weight and lower systolic blood pressure. At age 17-18 the seasonal fluctuations in systolic blood pressure have disappeared among males. Among females of the same age they still exist but are not correlated with the pattern in birth weight. With regard to the season-of-birth fluctuations our study therefore does not find any support for a negative relationship between birth weight and current systolic blood pressure.

Birth weight, and systolic blood pressure at age 11-12 in the twins and total mortality (males) and cancer (females) in the US death records follow similar bimodal fluctuations. This suggests that the factors that caused the summer trough and autumn peak in the life spans of people born at the beginning of the 20th century could still cause differences in birth weights and systolic blood pressure among children at the end of the 20th century.

Table 8.5. Previous studies about season-of-birth patterns in birth weight, current weight, current height, and weight/height gain.

Topic	Authors	Country	Time period	Size of the sample	Minimum	Maximum	Peak-to-trough difference
Birth weight by month of birth	Adair & Pollit 1983, 1985	Taiwan	6-year time period	450	JUN-AUG (rainy season)	NOV-APR (mild winter)	152 g
	Bantje & Niemeyer 1984	Tanzania	1972-1981	7134	FEB-MAR (rainy season) JUN	SEP-NOV	37 g
	Fallis & Hilditch 1989	rural Zaire (Z) Ontario, Canada (C)	JAN 1972–31 DEC 1981 (Z), 1979–1983 (C)	8,815 (Z); 15,954 (C)	NOV (Z) (rainy season); no seasonality in (C)	JUN (Z) (dry season); no seasonality in (C)	(Z) 196 g
	Hellmuth & Wnorowski 1923	Hamburg, Germany	JAN 1912–SEP 1922	26,515	DEC-MAR (winter)	JUN-OCT	
	Kinabo 1993	Morogoro, Tanzania	1985–1989	19,783	FEB-MAR (rainy season)	AUG (dry season)	
	Matsuda et al. 1995	Japan	JAN 1974–DEC 1983	7,622,012	summer and winter	spring and autumn	
	Moore 1998	The Gambia	1949–cont.	3,102	JUL-DEC (hungry season)	JAN-JUN (harvest season)	200–300 g
	Otto & Glass 1959	Leipzig, Germany	1936–1956	53,638	spring	autumn	43 g

Table 8.5. (continued)

Birth weight by month of birth	Otto & Noack 1957	Berlin; Leipzig; Marburg, Wien, Odessa; Germany	1936–1953, 1937–1953, 1920–1922	48,497 (B) 52,947 (L) 39,000 (M, W, O)	DEC–JUN (B, L); NOV–JUN (M, W, O)	JUL–NOV (B, L); JUL–OCT (M, W, O)	(B, L) 35 g (M, W, O) 60 g
	Pollitt & Arthur 1989	Sui Lin, Taiwan		225	MAY-AUG (hot, rainy season), summer	FEB-MAY (dry season), spring	218.7 g
	Rousham & Gracey 1998	Kimberley region, Western Australia	1981–1993	4,508	JAN–MAR, (wet season))	JUL–OCT, (dry season)	
	Ward 1987	Vienna, Austria	1865–1930	12,700	winter and spring	SEP-DEC autumn	58 g
	Wendl-Richter 1997	North Western Burkina Faso	1987–1989	2,026	AUG (hunger season, field work)	APR-MAY (dry season)	
Length at birth by month of birth	Otto & Glass 1959	Leipzig, Germany	1936–1956	53,638	winter	SEP-OCT	1.92 mm
	Otto & Noack 1957	Berlin & Leipzig, Germany	1936/37–1953	101444 (48,497 + 52,947)	DEC, JAN; MAY-JUL	MAR–APR, OCT	1.5 mm
	Wohlfahrt et al. 1998	Denmark	1973–1994	1,166,206	DEC	APR	2.2 mm

Table 8.5. (continued)

Topic	Authors	Country	Time period	Size of the sample	Minimum	Maximum	Peak-to-trough difference
Weight gain in infants by current month	Brown et al. 1982	Bangladesh	APR 1978–JUN 1979	174	MAR–JUN	AUG–DEC	
	Xu et al. 2001	Shanghai, China	JAN 1980–DEC 1990	4,128	MAY-JUL (wet season), summer	AUG-JAN autumn and winter	
Height gain in infants by current month	Cole 1993	West Kiang, The Gambia; Cambridge, England	MAY1974–APR 1980 (G), 1984 -1988 (E)	686 (G), 191 (E)	JUN–OCT (hungry season) (G)	AUG (E)	
	Xu et al. 2001	Shanghai, China	JAN 1980–DEC 1990	4,128	autumn and winter	spring and summer	
Weight loss by current month	Cole 1993	The Gambia females	1978–1988	529	JUN-OCT (hungry season)	NOV-MAY (harvest season)	4 kg in 1978 2 kg in 1988
	Chen et al. 1979	Bangladesh	MAR 1976-SEP 1977	200	MAY-JUL	SEP-OCT	2 kg
Current weight by month of birth	Chenoweth and Canning 1941	Cincinnati, United States	birth cohorts 1904–1921	10,005 freshmen	AUG–NOV (SEP–DEC)	MAR–JUN (spring)	0.8 pounds
	Fitt 1955	New Zealand	army draft World War II.	21,177 men	MAY – JUN (late autumn)	DEC (early summer)	1.32 pounds
	Hillman & Conway 1972	United States	1966–1970	9,103	normal & underweight: JUN and JUL; overweight: OCT and NOV	overweight: MAR	

Table 8.5. (continued)

Current weight by month of birth	Mills 1955	Kentucky, Kansas, Wisconsin, North Carolina	1912–1939	45000 freshmen and women (ages 16–19)	OCT–DEC	APR–JUN	boys: 0.71 pounds; girls: 0.41 pounds
	Phillips and Young 2000	Hertfordshire, UK	1920–1930	1,750	JUL–DEC	JAN–JUN	
Current height by month of birth	Chenoweth and Canning 1941	Cincinnati, United States	1904–1921	10,005 freshmen	AUG–OCT	MAR-JUN spring	0.3 inch
	Mills 1955	Kentucky, Kansas, Wisconsin, North Carolina	1912–1939	45,000	OCT–DEC	APR–JUN	boys: 0.16 inches, girls: 0.04 inches
	Weber et al. 1998	Austria	1984–1993 (birth cohorts 1966–1975)	507,125 men	autumn	spring	0.6 cm
	Banegas et al. (2001)	Spain		2021, ages 35-64	males: winter	males: summer	1.7cm
Current systolic blood pressure	Banegas et al (2000)	Spain		572 men ages 45-64	MAY-JUN	OCT-NOV	11.1 mmHG

9 Summary

9.1 Early-Life Circumstances and Adult Mortality

When mortality started to decline in the second half of the 19th century, the decline followed a cohort pattern, which is direct evidence for the effect of early-life circumstances on adult mortality. Early studies by Kermack et al. in 1934 suggest that, as far as mortality up to the year 1925 is concerned, the year of birth had more predictive power than the year of death. This was reflected in the general view of public-health policy at the beginning of the 20th century. Mortality at adult ages was considered to be largely influenced by conditions that existed during childhood and in the earliest periods of life. After 1925, the year of birth as compared to the year of death lost much of its predictive power. Period factors became more important.

Today there is widespread evidence that, during a mortality regime that is determined by infectious disease, early-life conditions play a detrimental role for adult health and mortality. One major cause of death that declined on a cohort basis rather than on a period basis during the last 100 years is tuberculosis. Tuberculosis in late adulthood is believed to originate in infections experienced during childhood. Other important causes of death that have exhibited a cohort decline since the beginning of the 20th century are stroke, stomach cancer, and bronchitis. In the case of peptic ulcer-morbidity, the cohort-specific decline can be traced back to cohort-specific Heligopacter pylori infections in childhood. It seems that heligobacter pylori infections in childhood caused by poor living circumstances before and during World War II are responsible for the high rates of stomach cancer in Japan today. Adult mortality in Sweden during the 18th and 19th centuries was heavily influenced by the disease load during the first year of life: people born in years with extremely high infant mortality due to smallpox and whooping cough had increased mortality later in life.

Much controversy exists, however, about whether cohort effects, and thus early-life circumstances, are still important in a mortality regime that is largely determined by chronic diseases. With the epidemiological transition from infectious to chronic disease, cohort effects on mortality have largely lost their importance. Mortality is determined by period effects.

Coronary heart disease initially shows period-specific increase; the introduction of antibiotics influenced mortality in a period-specific way; and medical therapeutics usually acts in a period-specific manner. One counterexample is lung cancer, which increased from cohort to cohort, reflecting the increase in the number of smokers among the younger cohorts. For this reason, the main emphasis in research and public policy shifted to life-course factors such as adult socioeconomic status and lifestyle factors such as diet, physical exercise, and smoking behavior.

In the 1970s Forsdahl, a medical doctor in Norway, observed that regional differences in adult lung cancer and heart disease were not related to contemporary differences in lifestyle, smoking behaviour, or socioeconomic status but rather to differences in regional infant mortality during childhood and youth of the cohorts under study. His study is now considered to be the starting point of a large and productive area of research that tries to link early-life conditions to the manifestation of chronic disease later in life. The discussion about early-life effects on health at adult ages gained momentum with studies conducted by the Southampton group of Barker and colleagues. The group developed the fetal-origins hypothesis of adult disease (also known as the 'Barker hypothesis'), which suggests that coronary heart disease at adult ages results from poor conditions *in-utero* caused by inadequate nutrition on the part of the mother and infectious diseases she suffered during pregnancy. Since inadequate nutrition of the fetus is reflected in low birth weight, the Barker hypothesis claims that growth retardation *in-utero* leads to low birth weight and to an increased risk of chronic disease later in life. It seems that the main connection between birth weight and heart disease later in life is systolic blood pressure – infants with a low birth weight experience increased systolic blood pressure at adult ages.

The fetal-origins hypothesis has led to a large amount of research which generally comes to the conclusion that low birth weight is associated with an increased risk of heart disease at adult ages and that low-birth weight infants suffer from increased systolic blood pressure later in life. The interpretation of these outcomes has been repeatedly challenged, however. The main idea underlying the fetal-origins hypothesis is that a critical time period exists early in life and that negative effects during this period cannot be reversed later in life. Critics of the hypothesis frequently bring forward the argument that birth weight is confounded with socioeconomic status. Negative social factors in the early-life environment may set people onto life trajectories that negatively affect their health over the whole life course. Therefore, the almost universally observed relationship between birth weight and the risk of chronic disease later in life may be an outcome of the whole life course rather than the result of a critical period early in life.

This criticism leads to the question whether one can find an indicator for the prenatal and early postnatal environment that is not related to the life-course. Birth weight certainly does not fulfil this criterion and, in addition, it is not widely available. Studies that use birth weight or other direct indicators of early-life circumstances are usually based on hospital data, which are invariable subject to selection bias. Moreover, their sample sizes tend to be modest. The fetal-origins hypothesis suggests that nutrition and infectious diseases during the pregnancy of the mother are responsible for growth retardation in the infant, which leads to an increased risk of heart disease at adult ages. Both nutrition and infectious diseases are highly seasonal: Respiratory infections peak in the autumn and winter, and gastrointestinal infections peak during warm periods of the summer months. The availability of fresh fruits and vegetables – and thus of micronutrients – changes with the seasons of the year. An indicator that reflects the seasonally changing environment during the prenatal and early postnatal period is month of birth.

The month of birth can be used as an indicator for the seasonally changing environment around the time of birth provided that it is not confounded by social factors and that it does not have any life-course interpretation. This monograph tests a series of alternative hypotheses which all assume the existence of life-course factors or social factors that are related to the month of birth. Only after having rejected all these hypotheses is a thorough analysis of the possible causal mechanisms performed.

Epidemiological research on the underlying factors of schizophrenia has long used month of birth as an indicator for early-life circumstances that affect the risk of schizophrenia later in life. This line of research dates back to Ellsworth Huntington, who in 1938 published his book *Seasonality*, in which he describes the relationship between the seasons of the year and social, psychological, and demographic phenomena. By 1997 more than 250 studies about the month-of-birth effect in Schizophrenia had appeared and many more are still being published. The most recent research links the relationship between the month of birth and the risk of schizophrenia to influenza infections experienced by the mother during pregnancy, although results are not always conclusive. A month-of-birth effect has also been shown for other mental disorders such as manic depression and bipolar disorder, autistic disorder, Alzheimer's disease or anorexia nervosa. Little research, however, has been done on the month-of-birth effect in chronic diseases, which are what determine life expectancy in contemporary populations.

9.2 How to Study the Effect of the Month of Birth on the Life Span?

This monograph establishes the month-of-birth effect in the life span on the basis of large and complete population data for over 17 million people. It is based mainly on the analysis of three populations in the Northern Hemisphere – Austria, Denmark and the United States – and one population in the Southern Hemisphere, namely Australia. The rationale underlying this selection of populations is that those in the Northern Hemisphere should reveal a similar month-of-birth pattern, while the pattern should be shifted by half a year in the Southern Hemisphere population.

For Austria, Australia, and the United States death records form the basis of the analysis. These death records contain the date of birth and the date of death. Since only deaths are known while the population at risk is unknown, mean age at death is used as a substitute for life expectancy. An advantage of using US data is that the US territory ranges over more than 20 degrees latitude and consists of six major climatic zones. Since the death records contain the state of birth of the decedents, the data allow for regional analysis. The Danish data come from the Danish population register. The register allows us to follow each person alive in March 1968 until the present and records date of birth, date of death for those who have died, and date at the end of the follow-up for those who survived. From these data both exposures (number of person months of people born in a specific month) and occurrences (number of deaths for people born in a specific month) are known. We thus estimate death rates and life expectancies based on life-table techniques.

The advantage of studying Austria and Denmark is that they are geographically close and should therefore reveal similar month-of-birth patterns. And although they are geographically close they belong to different climatic zones and have different histories of adult and infant mortality. At the beginning of the 20th century infant mortality in Denmark was about half the mortality in Austria, where short periods of breastfeeding – or the complete absence thereof – took its toll. Thus, despite their geographical proximity, early-life circumstances, in terms of disease environment for example, differed widely.

The data of all four populations are used to study the mean age at death for ages 50+ (Austria, Australia, and United States) and remaining life expectancy at age 50 (Denmark). Although age 50 is certainly an arbitrary cut-off point, there are several reasons why this age has been chosen. First, infant mortality is deliberately excluded in order to guarantee that differences in mean age at death are caused not by differences in survival during the first year of life but rather by differential survival later in life. Second, in contemporary populations mortality during childhood and young adulthood is so low that dividing populations according to their month of birth would result in even higher mortality fluctuations that may have distorted the general month-of-birth pattern. Third, mortality starts to accelerate around age 50, with the increasing incidence of chronic diseases. Thus, the month-of-birth pattern in longevity is statistically primarily a function of the number of deaths at middle and old ages.

The cut-point at age 50 is disadvantageous, however, in terms of information about the month-of-birth pattern at younger ages. To overcome this limitation three additional data sources are studied. Aggregate data on infant mortality in the years 1911-1915 in Denmark by age and month of death are used to estimate infant mortality in the first year of life by month of birth. The Danish twin registry allows for the study of all twins born between 1870 and 1930 who survived at least up to the age of six. The twins are followed up to the present, and for the deceased the date of death is recorded. Finally, the US census samples for the years 1960, 1970, and 1980 contain information about the quarter of birth. The comparison of the age-specific percentages of cohorts born in a specific quarter between two succeeding census rounds provides an indication of differential survival by quarter of birth. For example, one can compare the percentage of people born in the second quarter at ages 10 in the 1960 census with the respective percentage of the 20 year-olds in the 1970 census and the 30 year-olds in the 1980 census. If the percentage changes from one census to the next, then the quarter of birth influences survival. The direction of the change indicates whether those born in the second quarter experience higher or lower mortality as compared to the total population.

There is an additional advantage of the US census data and the Danish twin data. Both datasets constitute a second, independent source for the two populations that can be used to verify the month-of-birth patterns. Since the US results (together with the Austrian and Australian results) are based on death counts only, it could be argued that the result is at worst an artefact and at best confounded because of the approximation of life expectancy by mean age at death. The use of repeated census rounds avoids the calculation of life expectancy and is therefore not subject to this criticism. Both the Danish register data and the Danish twin data are longitudinal data and thus provide all the necessary information to calculate life expectancy. The advantage of the Danish twin register is that the month-of-birth pattern of the general population can be replicated in a very specific sub-population. Furthermore, one would expect that the peak-to-trough difference in life expectancy is larger among twins than among the general population because they experience harsher conditions early in life than singletons.

For the United States and Australia neither register data nor information about the seasonal distribution of birth dates at the beginning of the 20th century is available. Mean age at death is therefore used as an approximation of lifespan. This monograph shows that in two populations, the Danish twins and the Austrian general population, using mean age at death from death records of not extinct cohorts results into a minor bias concerning the month-of-birth pattern. For populations without information about the seasonal distribution of births and without population registries mean age at death gives therefore a reasonable approximation of the month-of-birth pattern.

9.3 The Month-of-Birth Effect Depends on the Season

A significant effect of the season of birth on survival at adult ages exists in all populations studied. In the Northern Hemisphere mortality at ages 50+ peaks for the spring-born (April–June) and reaches a trough for the autumn-born (October–December). In the Southern Hemisphere the pattern is shifted by half a year: mean age at death peaks for the spring-born and reaches a minimum for the autumn-born. The difference in life span between the spring- and the autumn-born is 0.6 years in Austria – twice that of Denmark (0.3 years). For the total US population the peak-to-trough difference is 0.4 years. There is, however, a large amount of variability by region of birth. With 0.31 years, the difference is lowest for people born in New England, followed by the Middle Atlantic region (0.36 years). It reaches a maximum for those born in the East South Central (0.86 years) and West South Central regions (0.69 years), while all other regions show differences of around 0.44 years. The regional variability in the peak-to-trough differences persists when the results are corrected for the effects of race, education, and marital status. The maximum difference in Australia, which is 0.6 years, compares to that of Austria, although the pattern is shifted by half a year, as noted above.

Similarly to the US death records, the Australian death data contain state and territory of birth. In general, the pattern is similar in all states and territories – with high mortality for decedents born in the fourth quarter and low mortality for those born in the second quarter. Within Australia the peak in mortality of those born in the fourth quarter is highest for decedents born in Victoria (0.89 years), followed by those born in South Australia (0.61 years), Western Australia (0.49 years) and in New South Wales (0.39 years).

There are two exceptions: the first is Tasmania, where the peak-to-trough difference is not significant due to the small number of decedents. The second exception is Queensland, which exhibits a pattern that differs from those of all the other states. Here the trough in mean age at death occurs for people born in the first four months, while the peak is for those born in April and May. Queensland belongs to tropical northern Australia and the trough in life expectancy coincides with the hot and wet summer season, the peak with the beginning of the more moderate and dry winter season. Since the US death records also contain information about decedents born in Hawaii, it is possible to compare the pattern of Queensland with that of Hawaii. The two regions belong to similar climatic zones. It turns out that the trough in life expectancy is linked to the hot summer period and the peak to the more moderate winter period also in Hawaii.

The results for the United States and Denmark are supported by the analysis of the US census rounds and the Danish twin register. The three US census rounds show a significant disadvantage in survival for people born in the second quarter and an advantage for those born in the autumn. Up to the age of 40, the conditional survival probability of males surviving the next 20 years is about one per cent higher for the autumn-born than for the general population; between the ages 40 and 60, it is about 2.4 percent; and it increases to 7.8 percent at ages 60–69.

In general, the Danish twins experience a month-of-birth pattern that is similar to that of the general Danish population. There are differences, however. The most important one is that the peak-to-trough difference in life expectancy is much larger among for twins than for the general population. On the basis of the Danish register data, the difference in remaining life expectancy at age 50 between the autumn- and the spring-born is 0.3 years; for the Danish twins it ranges from 0.58 years in the youngest cohort (1910–1930) to 0.88 years in the cohort 1890–1909. In other words, the difference is at least twice to three times as high for twins than for the general population. This result is in agreement with the initially formulated hypothesis that the harsher environmental circumstances *in-utero* and in the first year of life will result in larger differences in life span by month of birth. It must be noted, however, that the mortality advantage of the autumn-born is not as consistent for the male Danish twins as it is for the general Danish male population.

The results from Austria, Denmark, the United States, Australia, and Hawaii suggest that the month-of-birth pattern in lifespan is linked to the seasons of the year. It appears that those born during periods of moderate weather, which coincides with autumn for populations living at higher latitudes and winter in populations close to the equator, experience higher life expectancy. Those born during spring and early summer exhibit lower life expectancies; at lower latitudes, periods with high temperatures seem to be detrimental. At first glance it seems surprising that the month-of-birth pattern exists at all in regions close to the equator. Although it is commonly thought that there are no seasonal changes at lower latitudes, this is not true. Differences in temperatures, humidity, and precipitation define the seasons of the year in these areas. The total effect of these factors combined, however, may vary depending on the climate zone.

9.4 The Month-of-Birth Pattern of Migrants

A further test of whether the month-of-birth pattern depends on the seasonal environment early in life is to investigate the pattern of migrants. The Australian death records contain information on about 43,000 British-born immigrants to Australia who died at ages 50+. One would expect that these immigrants reveal the month-of-birth pattern of the Northern Hemisphere rather than that of the Southern Hemisphere. Since the peak in life expectancy for the November-to-January-born is 0.36 years higher among the immigrants than among the Australian-born while it is 0.26 years lower for the March-to-May-born, this seems indeed to be the case. Contrary to the unimodal patterns of the populations of Northern Hemisphere, however, the month-of-birth pattern of the migrants is bimodal. One possible explanation for this bimodal pattern is that the month-of-birth pattern of the total British population is bimodal. Another explanation is that migrants are not representative of the total British population. At present, no published information exists that would enable one to test the first explanation. An exploratory analysis of the month-of-birth effect on the basis of the genealogy of the British peerage did not yield any conclusive results.

This monograph provides two examples based on US death records that migrants are selected – both socially and as concerns health – and therefore not representative of the total population. The first example shows that social status and race influence the month-of-birth pattern in life span. The peak-to-trough difference in mean age at death among US decedents with a low level of education is, with 0.62 years, almost twice as large as among those with a high level of education (0.38 years), while the basic pattern remains unchanged. This is different in the case of race. Among African-American decedents the peak-to-trough difference is only slightly larger (0.57 years) than among white decedents (0.49 years). The shape of the pattern differs widely, however. Both for whites and African Americans, those born in June and July show the lowest mean ages at death, while the peak in life expectancy is shifted from autumn to winter among African-Americans. Those born in the autumn do not experience the same mortality advantage as the white population, although a secondary peak in mean age at death does exist.

An indication that health status affects the month-of-birth pattern is the larger peak-to-trough difference for never-married decedents (0.62 years) as compared to those who married (0.40 years). It is now generally agreed upon that the beneficial effect of marriage on health and mortality results from two forces. On the one hand, marriage has a protective effect. On the other hand, individuals are selected into marriage according to their health status, and frail individuals tend to remain unmarried. Thus, health status precedes marriage, and to a certain extent marital status can be regarded as an indicator of health.

The US death certificates thus provide evidence that social and health factors modulate the month-of-birth pattern in life span and that migrants, who are selected both socially and as regards health, may therefore reveal a pattern that is somewhat different from that of their population of origin.

The second example applies to internal US migration. It shows that the month-of-birth pattern of migrants differs from that of their population of origin. In the US, large disparities by region of birth exist in the peak-to-trough difference in mean age at death. The lowest difference exists for decedents born in New England, and the largest for those born in the South East, which is generally referred to as the South of the US. Due to the sizeable migration flows within the US, it is possible to study migrants who were born in the North and who died in the South and vice versa. The two migration flows differ remarkably in their social composition as well as in their timing of migration. North-to-South migration is primarily retirement migration of healthy, married, and comparably well-off Americans who moved to the "sunbelt". This migration flow has become increasingly popular during the last few decades. In contrast, South-to-North migration was primarily labour migration caused by the economic upheavals of the first part of the 20th century. This migration peaked in the 1960s. These migrants had usually been born in rural areas and had lost their land or their status as sharecroppers because of the increasing industrialization of the southern agricultural system.

It appears that the North-to-South migration of the retirees is consistent with the expectation that migrants reveal the month-of-birth pattern of their birth region rather than that of their residence region. The peak-to-trough difference of the migrants born in the North is smaller than it is for those who were born and who died in the South. This is not generally true for labour migrants from the South to the North. Although most of the southern migrants stem from rural areas, they are by no means a homogenous group. The social-class relations in the South were structured along the lines of land ownership, and those who did not own their land were particularly badly off. The analysis of the month-of-birth pattern of farmers and farm workers shows that, in the South, the peak-to-trough difference of the farm workers is significantly larger than that of farmers. In other words, the labour migrants from the South to the North probably are less representative of the general southern population than the retirement migrants from the North to the South.

Three conclusions can be drawn from the analysis of the month-of-birth patterns of the Northern and Southern Hemisphere and of the migrants between the hemispheres: First, the differences in life span by season of birth are linked to seasonal factors existing at the time of birth. Since the seasons in the Northern and the Southern Hemisphere are shifted by half a year the month-of-birth pattern is also shifted by half a year. Second, at higher latitudes with four distinct seasons, those born in the spring have a lower life expectancy than those born in the autumn. At lower latitudes close to the equator, those born during the hottest season of the year experience a lower life expectancy than those born during the moderate winter season. Third, the peak-to-trough difference in the month-of-birth pattern varies for different social groups: the higher the social status, the lower the difference in life span. There are two alternative implications of this result. Since the education level of the deceased is closely linked to the social status at the time of birth, the educational gradient in the peak-to-trough difference may reveal the impact of social status. Alternatively, educational status later in life may modify the effect of the early-life environment. This explanation implies that early life is a critical period in terms of adult health and mortality, although its effect may be partly reversed by life-course factors.

9.5 Alternative Hypotheses for the Month-of-Birth Effect

Before proceeding further with the analysis of the month-of-birth effect on the life span it is worthwhile to consider whether explanations exist other than those that are related to the beginning of life. One frequently raised concern is that the month-of-birth effect reflects the seasonal distribution of deaths rather than the seasonal changes in the early-life environment. More specifically, the concern is that the interaction between the seasonal distribution of deaths and the monthly increase in adult mortality causes a month-of-birth pattern. This hypothesis has been already widely discussed in the research about the month-of-birth effect in schizophrenia, whose incidence is seasonal and whose risk increases with age.

The Danish register data used in this study allow us to test this hypothesis explicitly due to the longitudinal nature of the dataset. The result is that, although month of birth, age, and month of death influence mortality simultaneously, they are independent of each other. This is consistent with the outcomes of the tests that were developed in the area of schizophrenia.

A second, frequently raised concern is that the month-of-birth effect is caused by socioeconomic differences in the seasonal distribution of births. The number of births is distributed seasonally over the year with the exception of only a few populations. If the seasonality in births is partly driven by the preference of couples for giving birth in certain seasons of the year, then this preference may differ between social groups. In schizophrenia research, this explanation is generally known as the procreational habits theory. Indviduals with schizophrenia may have a procreational pattern that differs from those of the non-schizophrenic population. Since schizophrenia is heritable to a large degree, this would result in a seasonal distribution of births of individuals at increased risk of developing schizophrenia that differs from the seasonal distribution in the general population. Two predictions of this theory have already been falsified: First, it is not the case that individuals with schizophrenia who have a family history of schizophrenia reveal a larger peak-to-trough difference in their seasonality of births than those who do not have a family history. Second, schizophrenic individuals and their siblings have different seasonal birth distributions.

In this study education is used as an indicator for social groups and the seasonal birth distribution at age 50+ is analysed on the basis of the 1981 census for Austria. It turns out that proportionally more people with a high level of education are born in spring than in the autumn, while proportionally more people with basic education have their birthday in the autumn. Among the cohorts analysed, one's own education is closely linked with the education of one's parents, so this result indicates that social differences in the seasonal distribution of births cannot explain the month-of-birth pattern in the life span. On the contrary, it appears that they even tend to reduce the differences in life span by month of birth.

Another prominent hypothesis in social and psychological research is the deadline hypothesis. Starting school is usually tied to reaching a certain age before a certain deadline. Children who are born shortly after the deadline have to wait an additional year before the can start school and will therefore be among the oldest of their classmates. This may pose a special advantage compared to those who are born shortly before the deadline, who will thus always be among the youngest. There is, however, also a second interpretation of the deadline hypothesis. Children born shortly after the school attendance deadline will be older when they start school and will attain the legal drop-out rate earlier in the educational year than those born shortly before the deadline. This may result in fewer years of school enrolment. Being born after the deadline is therefore considered to be disadvantageous in terms of one's educational career and research suggests that this is reflected in income later in life. The seasonal distribution of the month of birth of Austrians aged 50+ seems to provide support the second interpretation of the deadline hypothesis. In Austria, school started on 1 October, and children who had not turned six by that date had to wait another year before they could start school. Since the autumn-born belong to a larger percentage to the group with basic education only, the deadline effect would be detrimental rather than beneficial for those born shortly after the deadline of 1 October. However, since the mean age at death of the autumn-born is higher than that of the spring-born, the deadline hypothesis cannot explain the month-of-birth effect on the life span.

9.6 The Causal Mechanisms

What are the mechanisms that cause the month of birth pattern in the life span? One can rule out the possibility that the month of birth has any major life-course interpretation in terms of longevity. This means that any month-of-birth pattern must necessarily relate back to the very first period of life. One can also rule out the possibility that the month of birth is confounded by socioeconomic status and that the differences in life span by month of birth are simply differences in socioeconomic status.

The study of the month of birth pattern in infant mortality at the beginning of the 20th century allows us to rule out a third alternative as well. If autumn-born infants suffer higher mortality in their first year of life than spring-born infants, then the increased life expectancy of the autumn-born may simply be the result of selective survival. Among the autumn-born only the more robust individuals would survive the first year of life, who would then have lower mortality later in life. Danish data on historical infant mortality between the years 1911 and 1915 show, however, that it is the spring-born rather than the autumn-born who experience higher mortality in their first year of life. The standardized death rate of the June-born infants is 30 per cent higher than the death rate of the December-born. It thus appears that selective survival cannot explain the month-of-birth pattern in the life span.

The month-of-birth pattern in infant mortality points to another explanation. For Denmark the correlation between infant mortality in the first year of life and adult mortality after age 50 is 0.87 and highly significant. This result indicates that the causal mechanism of the month-of-birth pattern is related to debilitating factors *in-utero* or in the first year of life. In other words those factors that contributed to the high infant mortality of the past are also the factors that cause the differences in life span by month of birth. Public health experts at the beginning of the 20th century felt that mothers' health status and whether mothers breastfed their babies were the two most important factors, followed by housing, sanitation and general poverty, all of which were closely linked with the survival of infants. The health status of pregnant women depends largely on their diet and on the general disease load. Breastfeeding the infant is related primarily to a lower incidence of infectious diseases of the gastrointestinal tract, which historically is the major cause of infant mortality.

Both nutrition and infectious diseases are highly seasonal. Diet at the beginning of the 20th century did not much resemble contemporary dietary patterns. People ate less meat, fruits, and vegetables and more starchy staple food. The first vitamins were not discovered until 1911, and in the early 1900s nutritionists were even opposed to greens, which were considered to require more bodily energy for digestion that they provided. Although severe malnourishment was not widespread, people had inadequate nutrition – particularly during the winter and early spring. Peak growth of the fetus *in-utero* occurs during the third trimester. For infants born in spring, the third trimester coincides with a period of largely inadequate nutrition; for those born in the autumn it coincides with a period of plenty.

The incidence of infectious diseases depends on the climate and on the seasons of the year. The incidence of waterborne infectious diseases, which affect mainly the gastrointestinal tract, is correlated with warmer temperatures and flooding. Peak climatological temperatures coincide with the incidence of foodborne diseases. Many childhood diseases are highly seasonal; airborne diseases affecting the respiratory tract usually peak in autumn and winter.

The effect of the mother's nutrition during pregnancy on the health of the child at adult ages is highly contested. Nutrition is considered to be the central causal mechanism in the fetal-origins hypothesis, which claims that inadequate nutrition of the mother during pregnancy leads to growth retardation *in-utero* and to an increased risk of heart disease later in life. Studies that looked at the old-age mortality of cohorts born shortly after periods of famine, which were thus presumably marked by severe malnutrition of the mother during the gestational period of their unborn, did not find any differences. In other words, those born shortly before, during, or after a famine had similar mortality rates at older ages. On the other hand, the study of the social composition of births during and shortly after the Dutch Famine in 1945 has shown that this composition has changed. Primarily, women of higher social background continued to give birth while women of lower background stopped having babies. Unobserved differences in the social composition might therefore be the reason that cohorts born during and shortly after a famine do not have higher mortality later in life than those unaffected by the famine. Still, the evidence is weak concerning the effect of nutrition during gestational age on mortality later in life.

The situation is different concerning the effect of infectious diseases. There exists ample evidence that infectious diseases early in life affect later-life health and mortality. Historically, people born in years with extremely high infant mortality caused primarily by whooping cough and smallpox tend to have higher mortality later in life. Tuberculosis in late adulthood is believed to originate in infections during childhood. H pylori infections in childhood are responsible for peptic ulcer morbidity later in life. Increased H pylori infections in childhood caused by poor living circumstances before and during World Ware II seem to be responsible for the high rates of stomach cancer in Japan today. Chronic respiratory disease later in life seems to be related to lower respiratory tract infections in childhood.

The indicator "month of birth" does not allow us to distinguish between the period *in-utero* and the first year of life. It therefore does not tell us whether it is the nutrition of the mother during pregnancy or infectious diseases in the first year of life that are of crucial importance. The investigation of causes of death reveals that the month-of-birth effect exists for all major groups of causes of death. The pattern is particularly strong and consistent for heart disease. It is less strong but nevertheless consistent for many types of cancer. Significant month-of-birth patterns exist for causes of death such as stroke, stomach cancer, and bronchitis, which declined between the mid 19th century and today, following a birth cohort pattern. There are two possible explanations: First, since the etiology of all the different causes of death differs, there may be more than one causal mechanism and some of them may be related to the period *in-utero*, others to the first year of life. Research conducted by Barker and colleagues, who formulated the fetal-origins hypothesis, links the increased risk of heart disease later in life to nutrition *in-utero*. A large body of literature investigates the month-of-birth effect in the prevalence of schizophrenia and discusses the effect of influenza infection of the mother during pregnancy on the child's risk of developing schizophrenia. Historically, children born in years with extremely high infant mortality caused by smallpox experience higher mortality later in life. A second explanation is that the spring-born are generally weaker than the autumn-born and that their rate of aging is higher.

Explanations other than nutrition and infectious disease have also been brought forward to explain the month-of-birth effect. One of the first to study the influence of the month of birth on the life span is Elsworth Huntington, who formulated the hypothesis in 1936 that high temperatures at the time of conception weaken the "germ plasma" of the parents, with negative effects on the development of the foetus. Recent research has shown that the sperm quality of men who work outdoors does indeed decrease during periods of high temperatures. A related hypothesis is that hot summers are the cause of protein deficiencies at the time of conception. This hypothesis is clearly ruled out on the basis of the US death data. The United States consists of six major climatic zones with very different climatic conditions. Since the US death data contain the state of birth, it is possible to correlate the peak-to-trough difference in life span by month of birth for people born in a specific state with maximum and minimum temperature and with the maximum difference in temperature. It appears that no correlation exists between the peak-to-trough difference and the temperature variables, neither for total mortality nor for major causes of death.

Another explanation is that seasonal changes in the hours of daylight influence the human endocrine functions and that the month-of-birth effect might be caused by variations in the internal chemistry or neural development brought about by the seasonal variations in light. This hypothesis implies that the peak-to-trough differences should be larger at higher than at lower latitudes. Again, the hypothesis can be refuted on the basis of the US death data, since there is no correlation between latitude and the peak-to-trough difference in life span by month of birth.

9.7 Does the Month-of-Birth Effect Exist in Cohorts Born Today?

All the observed differences in life span by month of birth pertain to individuals who are born during the first part of the 20th century. The Danish results are based on the birth cohorts 1868 to 1918, the Austrian results on births between 1888 and 1946, the Australian results on births between 1893 and 1947, and the US results on births between 1889 and 1947. The question whether the differences in life span by month of birth still exist in infants born today cannot be answered on the basis of mortality data, since one does not know anything about the future mortality of contemporary cohorts. Other measures have to be applied.

The fetal-origins hypothesis of adult disease states that growth retardation *in-utero* leads to low birth weight and causes an increased susceptibility to chronic diseases at adult ages. According to Barker and colleagues, the main reason for growth retardation *in-utero* is deficient nutrition. Since nutrition is seasonal, one would expect that growth retardation is seasonal and that this is reflected in a seasonal pattern in birth weight. Evidence for seasonal growth retardation *in-utero* exists from less developed countries. A positive correlation between the seasonal pattern in birth weight and the month-of-birth pattern in the life span after age 50 of cohorts born earlier in time would suggest that the differences in life span still exist in contemporary cohorts.

This hypothesis is tested on twin data from the Minnesota twin registry. The seasonal pattern in the birth weight of twins born in the 1970s and 1990s is compared with the month-of-birth pattern in the mean age at death of decedents aged 50+ who were born in Minnesota. Twin data are used because twins belong to the most vulnerable group early in life in contemporary populations. They have higher spontaneous abortion rates and higher infant and child mortality up to the age of six. At adult ages, however, their mortality does not differ from the mortality of singletons. This is also true for the month-of-birth pattern in the life span, as has been shown in this monograph by the Danish twins and the Danish general population. There is, however, one difference between the Danish twin population and the general population. The peak-to-trough difference in the month-of-birth pattern is two to three times as large among the twins.

Since twins face much tougher conditions early in life than singletons, a seasonal pattern in birth weight due to growth retardation may be more readily detectable among twins. Results suggest that the seasonal pattern in birth weight is bimodal, with birth weight reaching its maximum in spring and autumn and its minimum in winter and summer. This result is consistent with earlier findings that the risk of pre-term births in Minnesota is bimodal, with peaks in summer and in winter.

A positive correlation of borderline significance exists between the life span pattern and the birth weight pattern for male twins. For female twins the results are not conclusive. This result may be a first indication that the differences in life span by month of birth still exist in cohorts born today.

Although the evidence presented is weak, one has to keep in mind that birth weight is only an imperfect measure of growth retardation since it does not take gestational age into account. For example, preterm infants may have just the right birth weight for their gestational age. Even if their birth weight is lower than that of term babies, they may not have suffered from growth retardation. A better measure would be birth weight corrected for gestational age, but information about gestational time is not available for the Minnesota twins. An earlier study of seasonal differences in the risk of preterm births in Minnesota shows that the risk is elevated during winter and summer. Thus, the seasonal pattern in birth weight of the Minnesota twins reflects not only seasonal growth retardation but also seasonal differences in gestational age. Further studies should therefore examine the correlation between the seasonal pattern in birth weight corrected for gestational age and the month-of-birth pattern after age 50.

The second "cornerstone" of the fetal-origins hypothesis is the claimed negative correlation between low birth weight and increased systolic blood pressure later in life. Growth retardation caused by deficient nutrition *in-utero* is reflected in low birth weight, and it leads to increased systolic blood pressure later in life and therefore to an increased risk of chronic diseases. The negative relationship between birth weight and systolic blood pressure has been found in a large series of studies, but has it also been challenged repeatedly. In terms of seasonal differences in nutrition, this hypothesis implies a) that a month-of-birth pattern exists in systolic blood pressure, and b) that the pattern is correlated with the seasonal pattern in birth weight. The Minnesota twin data also allow us to test this hypothesis.

In adolescence and young adulthood significant month-of-birth patterns exist for height, weight and systolic blood pressure, particularly among males. In adolescence about half of the week-of-birth pattern in systolic blood pressure is explained by the week-of-birth pattern in height and weight and the other half by the week-of-birth pattern in birth weight. This implies that, in adolescence, about half of the seasonal early-life effects on systolic blood pressure are related to factors *in-utero* that simultaneously affect birth weight and later systolic blood pressure. The other half affect systolic blood pressure through weight and height. The effect of birth weight is larger for males than for females.

With increasing age, the month-of-birth patterns in systolic blood pressure, weight, and height become less distinct, and birth weight contributes only marginally to the explanation of the season-of-birth pattern in systolic blood pressure. At that age, most of the explanatory power comes from the season-of-birth patterns in weight and height. This suggests that, at young adult ages, the season-of-birth differences in systolic blood pressure are related primarily to early-life factors that affect height and weight rather than birth weight.

9.8 Summary of Accomplishments and Outlook

This monograph has achieved six major accomplishments:

- First, it demonstrates the widespread existence of differences in life span by month of birth in the elderly populations of contemporary societies.
- Second, it provides evidence that the pattern is linked to the seasons of the year, by comparing the Northern and the Southern Hemisphere.
- Third, it formulates and tests a series of explanations for the month-of-birth effect and rejects many of the most frequently offered explanations. In particular, it rejects those that attribute the month-of-birth effect to social or statistical confounding.
- Fourth, in the case of Denmark it establishes a link between the month-of-birth pattern in the life span and the month-of-birth pattern in survival during the first year of life. The major determinants of infant mortality at the beginning of the 20th century are the health status of the mother and whether or not the child was breastfed. The first is related to the nutritional and disease environment of the mother during pregnancy, the latter to the risk to the infant of suffering from infectious diseases in the first year of life. The positive correlation between the month-of-birth pattern in the life span after the age of 50 and infant survival during the first year of life is therefore direct evidence that nutrition and infectious disease early in life play an important role in adult health and survival later in life.
- Fifth, it provides preliminary evidence that the differences in life span by month of birth still exist in cohorts born today.
- Finally, it contributes to the ongoing debate about the pathway that links growth retardation *in-utero* and the increased risk of chronic disease later in life by presenting evidence that this pathway need not necessarily be systolic blood pressure.

Although the month of birth has been used for a long time in research on mental disorders, the question as to whether a month-of-birth pattern exists in different populations and for different causes of death was never tested systematically. Research was mainly confined to specific diseases, historically this were mental diseases, more recently diseases such as diabetes, multiple sclerosis, or lung and breast cancer. Since the interest in the month-of-birth effect arouse primarily from an epidemiological point of view, more emphasis was given to the specific disease rather than to the phenomenon of the month-of-birth effect itself. Usually, the datasets are rather small and the results subject to random fluctuations. With some notable exceptions in the research on schizophrenia, anorexia nervosa, etc, the intention of most of the studies was to show that a month-of-birth effect exists rather than to test hypotheses regarding the underlying factors of the month-of-birth effect.

Some first steps to uncover the causal mechanisms behind the month-of birth effect in the life span are presented in this monograph. The next steps will have to consider the following topics: The indicator "month of birth" does not allow us to establish whether the critical period early in life is *in-utero* or the first year of life. It also does not provide an answer to the question whether it is infectious diseases or nutrition that cause the differences. One way to address this question is to study the effect of food shortage – and in its severest form, the effect of famine – on the differences in life span by month of birth. Peak growth *in-utero* takes place during the third trimester. Infants born in the spring experience worse nutritional conditions during the last trimester than infants born in the autumn. This is consistent with the finding that the spring-born have a lower life expectancy than the autumn-born. So the hypothesis is that foetuses affected by famine during their last trimester of gestation should experience higher mortality risks later in life.

Based on current research findings this line of research does not seem to be very promising, however. Studies of cohorts born during or shortly after a famine do not experience increased old-age mortality and the old-age mortality of twins is similar to that of singletons, although they usually have much lower birth weights.

The study of infectious diseases, on the other hand, shows more potential. There is ample evidence that infectious diseases during infancy and childhood affect late-life mortality. In Sweden, historical cohorts with exceptionally high infant mortality also have higher old-age mortality. In Denmark, the month-of-birth pattern in infant mortality at the beginning of the 20th century is positively correlated with the month-of-birth pattern in the life span at ages 50+ at the end of the 20th century.

To test whether infectious disease plays an important role, one can make use of the seasonal pattern of major outbreaks of infectious diseases. One example is the Spanish flu pandemic in 1918. Another possibility is the use of seasonal indices for the incidence of respiratory disease, gastrointestinal disease, and common childhood diseases such as smallpox, whooping cough, or rubella. One could test whether there is a positive correlation between these indices and the month-of-birth patterns for specific causes of death. Unfortunately, these indices are not readily available for the beginning of the 20th century. This means that contemporary indices would have to suffice, under the assumption that the seasonality has not changed over time. Another problem is that these indices are not available on an individual basis for the elderly population, so one would therefore have to use an ecological study design. A third possibility is to use seasonal infant mortality as an indicator of the infectious disease environment in the first three months, the second three months, and the second half-year. The three time periods should reflect the developmental stages of the infants regarding the immune system and the likelihood of weaning. A similar approach could be applied using seasonal adult mortality during childbearing ages to account for the disease environment of the mother during the three trimesters of pregnancy.

Future research into the month-of-birth effect on the life span and into the causes of death will have to proceed from the description of the phenomenon to finding the underlying causal mechanisms. They will probably differ for different causes of death. Some first steps to unravelling the mechanisms have been taken in this monograph, but much remains to be done.

References

Aalberse, R.C., Nieuwnhuys, E.J., Hey, M., Stapel, S.O. (1992): Horoscope Effect not Only for Seasonal but Also for Non-seasonal Allergens. Clinical & Experimental Allergy 22(11), 1003–1006

Aber, N. (1989) Birth Season Variation in Asthma and Allergic Rhinitis. Clinical Experimental Allergy 19, 643–648

Adair, L.S., Pollit, E. (1983): Seasonal Variation in Pre- and Postpartum Maternal Body Measurements and Infant Birthweights. American Journal of Physical Anthropology 62, 325–331

Adair, L.S., Pollit, E. (1985): Outcome of Maternal Nutritional Supplementation: A Comprehensive Review of the Bacon Chow Study. American Journal of Clinical Nutrition 41, 948–978

Alton, A., Massey, A. (1998): Date of Birth and Achievement in GCSE and GCE A-Level. Educational Research 40(1), 105–109

Angrist J.D., Krueger, A.B. (1991): Does Compulsory School Attendance Affect Schooling and Earnings?. The Quarterly Journal of Economics 106(4), 979–1014

Banegas, J.R., Rodriguez-Artalejo F., de la Cruz J.J., Graciani A., Villar Rey-Calero J. (2000) Adult Men Born in Spring Have Lower Blood Pressure. Journal of Hypertension 18(12), 1763–1766

Banegas, J.R., Rodriguez-Artalejo F., Graciani A., de la Cruz J.J., Gutierrez-Fisac, J.L. (2001) Month of Birth and Height of Spanish Middle-aged Men. Annals of Human Biology 28(1), 15–20

Bantje, H., Niemeyer, R. (1984): Rainfall and Birthweight Distribution in Rural Tanzania. Journal of Biosocial Science 16, 375–384

Barak, Y., Ring, A., Sulkes, J., Gabbay, U., Elizur, A. (1995): Season of Birth and Autistic Disorder in Israel. American Journal of Psychiatry 152(5), 798–800

Barker, D.J.P. (2002): Commentary: Components in the Interpretation of the High Mortality in the County of Finnmark. International Journal of Epidemiology 31, 309–310

Barker, D.J.P., Osmond, C. (1986a): Infant Mortality, Childhood Nutrition, and Ischemic Heart Disease in England and Wales. Lancet 10(1), 77–81

Barker, D.J.P., Osmond, C. (1986b): Childhood Respiratory Infection and Adult Chronic Bronchitis in England and Wales. British Medical Journal 293(6557), 1271–1275

Barker, D.J.P., Osmond, C. (1987): Death Rates from Stroke in England and Wales Predicted from Past Maternal Mortality. British Medical Journal 295 (6590), 83–86

Barker, D.J.P., Osmond, C., Golding, J., Kuh, D., Wadsworth M.E.J. (1989a): Growth In-utero, Blood Pressure in Childhood and Adult Life, and Mortality from Cardiovascular Disease. British Medical Journal 298(6673), 564–567

Barker, D.J.P., Osmond, C., Golding, J. (1990): Height and Mortality in the Countries of England and Wales. Annals of Human Biology 17(1), 1–6

Barker, D.J.P. (1995): Fetal Origins of Coronary Heart Disease. British Medical Journal 311, 171–174

Barker, D.J.P. (1994): Mothers, Babies and Diseases in Later Life. British Medical Journal Publishing Group, London

Bengtsson, T., Lindstrom, M. (2001): Childhood Misery and Disease in Later Life. The Effects on Mortality in Old Age of Hazards Experienced in Early Life, Southern Sweden, 1760–1894. Population Studies (3), 263–277

Ben-Shlomo, Y., Kuh, D. (2002): A Life Course Approach to Chronic Disease Epidemiology. Conceptual Models, Empirical Challenges and Interdisciplinary Perspectives. International Journal of Epidemiology 31, 285–293

Ben-Shlomo, Y., Davey Smith, G. (1991): Deprivation in Infancy or in Adult Life. Which is More Important for Mortality Risk?. Lancet 2, 530–534

Bernstein, L., Chilvers, C., Murrels, T. et al. (1986): Malignant Germ Cell Tumours of the Testis. Journal of Epidemiology and Community Health 40, 214–217

Biggar, J.C. (1980): Who Moved Among the Elderly, 1965–1970. A Comparison of Types of Older Movers. Research on Aging 2, 73–91

Bingham, C., Arbogast, G.C., Guillaume, L.J.K. (1982): Inferential Statistical Methods for Estimating and Comparing Cosinor Parameters. Chronobiologia 9, 397–439

Bleuler, M. (1991): Seasonality of Births of Future Schizophrenics and Seasonality of Depressive Episodes (letter) Schizophrenia Bulletin 17, 191–192

Block, T., Pedersen, C. R., Volund, A., Pallesen. C., Buschard, K. (1994): Perinatal Determinants Among Children Who Later Develop IDDM. Diabetes Care 17, 1154–1157

Bobak, M., Gjonca, A. (2001): The Seasonality of Live Birth is Strongly Influenced by Sociodemographic Factors. Human Reproduction 16(7), 1512–1517

Bolton, P., Pickles, A., Harrington, R., McDonald, H., Rutter, M. (1992): Season of Birth. Issues Approaches and Findings for Autism. Journal of Child Psychological Psychiatry 33, 509–530

Bound, J., Jaeger, D.A. (1996): On the Validity of Season of Birth as an Instrument in Wage Equations. A Comment on Angrist and Krüger's "Does Compulsory School Attendance Affect Schooling and Earnings. NBER Working Paper 5835, Cambridge, Massachusetts

Bradbury, J. H. Miller, G. A. (1985): Season of Birth in Schizophrenia. A Review of Evidence, Methodology, and Etiology. Psychological Bulletin 98, 569–594

Breitinger, F.E. (1966): Advances in the Biology of Human Populations 139–148, Hungarian Academy of Sciences, Budapest.

Breschi, M., Bacci, M.L. (1996): Month of Birth as a Factor in Children's Survival. In: Desjardins, B., Brignoli, P. (Eds.): Infant and Child Mortality in the Past. Clarendon Press, Oxford, 157–173

Brown, K.H., Black R.E., Becker S. (1982): Seasonal Changes in Nutritional Status and the Prevalence of Malnutrition in Longitudinal Study of Young Children in Rural Bangladesh. American Journal of Clinical Nutrition 36, 303–313

Brown, K.H., Black, R.E., Robertson, A.D., Becker, S. (1985): Effects of Season and Illness on the Dietary Intake of Weanlings During Longitudinal Studies in Rural Bangladesh. American Journal of Clinical Nutrition 41, 343–355

Brownlee, J. (1916): Certain Considerations Regarding the Epidemiology of Phthisis Pulmonalis. Public Health 29, 130–145

Brun, T. (1984): Physiological Measurement of Activity Among Adults Under Free Living Conditions. In: Pollit, E., Amante, P. (Eds.): Energy Intake and Activity. New York (Alan T. Liss, Inc.), 131–156

Carroll, R.E., Haddon, W. jr. (1964): Birth Characteristics of Persons Dying of Cerebral Aneurysms. Journal of Chronic Diseases 17, 705–711

Caselli G., Vallin J., Vaupel J.W., Yashin, A.I. (1985): Age-specific Mortality Trends in France and Italy Since 1900 – Period and Cohort Effects. European Journal of Population 3 (1), 33–60

Caselli G., Capocaccia R. (1989): Age-Period-Cohort and Early Mortality. An Analysis of Adult Mortality in Italy. Population Studies 43, 133–153

Caselli G. (1990): The Influence of Cohort Effects on Differentials and Trends in Mortality. In: Vallin, J.S., D'Souza, S., Palloni, A. (Eds): Measurement and Analysis of Mortality. New Approaches, Oxford, England, Clarendon Press, 229–249

Castrogiovanni, P., Iapichino, S., Pacchierotti, C., Pieraccini, F. (1998) Season of Birth in Psychiatry: A Review. Neuropsychobiology 37, 175–181

Centola, G., Eberley, S. (1999): Seasonal Variations and Age-Related Changes in Human Sperm Count, Motility, Motion Parameters, Morphology, and White Blood Cell Concentration. Fertility and Sterility 72, 803–808

Chambers, R., Longhurst, R., Pacey, A. (1981): Seasonal Dimensions to Rural Poverty. Frances Pinter Publishers, London

Chen, L.C., Chowdhury, A.K.M.A., Huffman, S.L. (1979): Seasonal Dimensions of Energy Protein Malnutrition in Rural Bangladesh. The Role of Agriculture, Dietary Malnutrition, and Infection. Ecology of Food and Nutrition 8, 175–87

Chenoweth, L.B., Canning, R.G. (1941): Relation of Season of Birth to Certain Attributes of Students. Human Biology 13, 533–540

Chew, F.T., Goh, D.Y.T., Teo, J., Quak, S.H., Lee, B.W. (1998): Month of Birth and Childhood Atopic Disease in the Tropics. Allergy 53(10), 962–968

Chotai, J., Salander Renberg, E., Jacobsson, L. (1999): Season of Birth Associated with the Age and Method of Suicide. Archives of Suicide Research 5(4), 245–254

Christensen, K., Orstavik, K.H., Vaupel, J.W. (2001): The X-Chromosome and the Female Survival Advantage. An Example of the Intersection Between Genetics, Epidemiology and Demography. Annals of the New York Academy of Sciences. 954, 175–183

Christensen, K., Stovring, H., McGue, M. (2001): Do Genetic Factors Contribute to the Association Between Birthweight and Blood Pressure. Journal of Epidemiology and Community Health 55, 583–587

Christensen, K., Wienke, A., Skytthe, A., Holm, N.V., Vaupel, J.W., Yashin, A.I. (2001): Cardiovascular Mortality in Twins and the Fetal Origins Hypothesis. Twin Research, 4(5), 344–349

Clark, R. (1993): Water: The International Crisis. MIT Press, Cambridge

Coale, A.J., Kisker, E.E. (1990): Defects in Data on Old Age Mortality in the United States. Asian and Pacific Population Forum 4(1), 1–31

Cole, T.J. (1993): Seasonal Effect on Physical Growth and Development. In: Ulijaszek, S.J., Strickland, S.S. (Eds.): Seasonality and Human Ecology. Society for the Study of Human Biology Symposium Series: 35, Cambridge University Press, Cambridge

Colwell, R.R. (1996): Global Climate and Infectious Disease. The Cholera Paradigm. Science 274, 2025–2031

Dahlquist, G.G., Källen, B.A.J. (1996): Time-spacing Clustering of Date of Birth in Childhood-onset Diabetes. Diabetes Care 19, 328–332

Dalen, P. (1975): Season of Birth. A Study of Schizophrenia and Other Mental Disorders. Amsterdam.

Datta, B.N.D. (1978): Epidemiology of Gastroenteritis of Preschool Children in Slum Areas in New Delhi with Special Reference to Helminth and Parasitic Infection. The Indian Journal of Pediatrics 45, 305–309

Davey Smith, G., Kuh, D. (2001): Commentary. William Ogilvy Kermack and the Childhood Origins of Adult Health and Disease. International Journal of Epidemiology 30, 696–703

Davey Smith, G., Harding, S., Rosato, M. (2000): Relation Between Infant's Birthweight and Mothers' Mortality. Prospective Observational Study. British Medical Journal 320, 839–840

Davey Smith, G., Hart, C.L., Ferrell, C., Upton, M.N., Hole, D.J., Hawthorne, V.M., Watt, G.C.M. (1997): Birthweight of Offspring and Mortality in the Renfrew and Paisley Study. Prospective Observational Study. British Medical Journal 315, 1189–1193

Davis, G., Welham, J., Chant, D., Torrey, E.F., McGrath, J. (2003) A Systematic Review and Meta-analysis of Northern Hemisphere Season of Birth Studies in Schizophrenia, Schizophrenia Bulletin 29(3), 587–593

De Montis, G. (1998): Circannual Rhythm for IgE-dependent Hypersensitivity to House Dust Mites According to Birthdate. Interpretation and Practical Consequences for Prevention. Bulletin del Academie Nationale de Medicine 182(8), 1697–1707

Derrick, V.P.A. (1927): Observations on (1) Errors in Age in the Population Statistics of England and Wales and (2) the Changes in Mortality indicated by the National Records. Journal of the Institute of Actuaries 58, 117–159

Dirks, R.T., Duran, N. (2001): African American Dietary Patterns at the Beginning of the 20th Century. Journal of Nutrition 131(7), 1881–1889

Doblhammer, G., Vaupel, J.W. (2001): Lifespan Depends on Month of Birth. Proceedings of the National Academy of Sciences of the United States of America. 98(5), 2934–2939

Doblhammer, G., Rodgers, J., Rau, R. (2000): Seasonality of Birth in Nineteenth and Twentieth Century Austria. Social Biology 47(3), 201–217

Dorling, D., Mitchell, R., Shaw, M. et al. (2000): The Ghost of Christmas Past. Health Effects of Poverty in London in 1896 and 1991. British Medical Journal 321(7276), 1547–1551

Dudink, A. (1994): Birth Date and Sporting Success. Nature 368, 592

Duffy, D.L., Mitchell, C.A. (1991): Respiratory Symptoms in Queensland and Australian Schoolchildren – an Association Between Month of Birth and Respiratory Illness. Clinical & Experimental Allergy 21(2), 231–234

Dyson, L.K. (2000): American Cuisine in the 20th Century. Food Review 23(1), 2–7

Eagles, J.M., Andrew, J.E., Johnston, M.I., Easton, E.A., Millar, H.R. (2001): Season of Birth in Females with Anorexia Nervosa in Northeast Scotland. The International Journal of Eating Disorders 30(2), 167–175

Ekbom, A., Daszak, P., Kraaz, W., Wakefield, A.J. (1996): Crohn's Disease After in-utero Measles Virus Exposure. Lancet 348, 515–517

Elford, J., Shaper, A.G., Whincup, P. (1992): Early Life Experience and Cardiovascular-Disease – Ecological Studies. Journal of Epidemiology and Community Health 46(1), 1–8

Elo, I.T., Preston, S.H. (1992): Effects of Early-Life Conditions on Adult Mortality: A Review. Population Index 58(2), 186–212

Elo, I.T., Preston, S.H. (1996): Educational Differentials in Mortality. United States, 1979–85. Social Science & Medicine, 42(1), 47–57

Eriksson, J.G., Forsen, T., Tuomilehtom, J., Osmond, C., Barker, D.J.P. (2000): Fetal and Childhood Growth and Hypertension in Adult Life. Hypertension 36, 790–794

Eriksson, J.G., Forsen, T., Tuomilehto, J., Osmond, C., Barker, D.J.P. (2001): Early Growth and Coronary Heart Disease in Later Life. Longitudinal Study. British Medical Journal 322, 949–953

Eriksson, J.G., Forsen, T., Tuomilehto, J., Winter, P.D., Osmond, C., Barker, D.J.P. (1999): Catch-up Growth in Childhood and Death from Coronary Heart Disease. Longitudinal Study. British Medical Journal 318, 427–31

Eriksson, N.E., Holmen, A. (1996): Skin Prick Tests with Standardized Extracts of Inhalant Allergenens in 7099 Adult Patients with Asthma or Rhinitis. Crosssensatizations and Relationships to Age, Sex, Month of Birth and Year of Testing. Journal of Investigational Allergology and Clinical Immunology 6(1), 36–46

Erel, F., Karaayvaz, M., Caliskaner, Z., Ozanguc, N. (1998): The Allergen Spectrum in Turkey and the Relationship Between Allergens and Age, Sex, Birth Month, Birthplace, Blood Groups and Family History of Atopy. Journal of Investigational Allergology and Clinical Immunology 8(4), 226–233

Eysenck, H.J., Nias, D.K.B (1982): Astrology. Science or Superstition? Temple Smith, London

Fallis, G. Hidditch, J. (1989): A Comparison of Seasonal Variation in Birthweights Between Rural Zaire and Ontario. Canadian Journal of Public Health 80 (3), 205–208

Feltbower, R., Pearce, M.S., Dickinson, H.O., Parker, L., McKinney, P.A. (2001): Seasonality of Birth for Cancer in Northern England, UK. Paedriatric and Perinatal Epidemiology 15, 338–345

Fichera, G., Arpi, M.L. Squatrito, S., Purello, F., Ashkenazi, I., Laron, Z. (2001): Seasonality of Birth in Children (0–14 Years) with Diabetes Mellitus Type 1 in the District of Catania. The Journal of Pediatric Endocrinology and Metabolism 14, 95–96

Fitt, A.B. (1955): The Heights and Weights of Men According to Month of Birth. Human Biology 27, 138–142

Fitt, A.B. (1941): Seasonal Influence on Growth, Function and Inheritance. New Zealand Council for Educational Research, Oxford University Press.

Fligstein, N. (1981): Going North. Migration of Blacks and Whites from the South, 1900–1950. Academic Press, New York

Floud, R., Wachter, K., Gregory, A. (1990): Height, Health and History. Cambridge University Press, Cambridge

Flynn, J.M., Rahbar, M.H., Bernstein, A.J. (1996): Is There an Association Between Season of Birth and Reading Disability?. Journal of Developmental and Behavioral Pediatrics 17(1), 22–26

Fogel, R.W. (1994): Economic Growth, Population Theory, and Physiology: the Bearing of Long-term Processes on the Making of Economic Policy. The American Economic Review 84, 369–395

Fogelman, Y., Rakover, Y., Luboshitzky, R. (1995): High Prevalence of Vitamin D Deficiency Among Ethiopian Women Immigrants to Israel: Exacerbation During Pregnancy and Lactation. Israel Journal of Medical Science 31(4), 221–224

Forsdahl, A. (2002): Observations Throwing Light on the High Mortality in the County of Finnmark. Is the High Mortality Today a Late Effect of Very Poor Living Conditions in Childhood and Adolescence?. International Journal of Epidemiology 31, 302–308

Forsdahl, A. (2002a): Commentary: Childhood Deprivation and Adult Mortality. International Journal of Epidemiology 31, 308

Forsdahl, A. (1973): Momenter til Belysning ar den Høye Dødelighet, Finnmark Fylke. Tidsskr Nor Lægeforen 93, 661–667

Forsdahl, A. (1977): Living Conditions in Childhood and Subsequent Development of Risk Factors for Ateriosclerotic Heart Disease. Journal of Epidemiology and Community Health 32, 34–37

Forsdahl, A. (1978): Are Poor Living Conditions in Childhood and Adolescence an Important Risk Factor for Arteriosclerotic Heart Disease. British Journal of Preventive and Social Medicine 31, 91–95

Forsen, T., Eriksson, J.G., Tuomilehto, J., Osmond, C., Barker, D.J.P. (1999): Growth In-utero and During Childhood Among Women who Develop Coronary Heart Disease. Longitudinal Study. British Medical Journal 319, 1403–1407

Forsen, T., Eriksson, J.G., Tuomilehto, J., Reunanen, A., Osmond, C., Barker, D.J.P. (2000): The Fetal and Childhood Growth of Persons who Develop Type 2 Diabetes. Annals of Internal Medicine 133, 176–182

Fossey, E., Shapiro, C.M. (1992): Seasonality in Psychiatry – A Review. Canadian Journal of Psychiatry – Revue Canadienne de Psychiatrie 37, 299–308

Frost, W.H. (1939): The Age Selection of Mortality from Tuberculosis in Successive Decades. American Journal of Hygiene 20, 91–96

Gavrilov, L.A. Gavrilova, N.S.(2003) Early-life Factors Modulating Lifespan, in Rattan, S.I.S (ed.), Modulating Aging and Longevity, Kluwer Academic Publishers, Dordrecht, 27–50

Gavrilova, N.S., Gavrilov, L.A., Evdokuskina, G.N., Semyonova V.G. (2002): Early Life Conditions and Later Sex Differences in Adult Life. Paper Presented at the PAA 2002, Atlanta

Goldberg, A.E., Newlin, D.B. (2000): Season of Birth and Substance Abuse. Findings from a Large National Sample. Alcoholism. Clinical and Environmental Research 24(6), 774–780

Goldman, N., Hu, Y. (1993): Excess Mortality Among the Unmarried. A Case Study of Japan. Social Science and Medicine 36(4), 533–59

Goldman, N., Lord, G., Hu, Y. (1993): Marriage Selection and Age Patterns of Mortality. A Mathematical Investigation. Mathematical Population Studies 4(1), 51–73

Greenwood, M. (1936): English Death-rates. Past, Present and Future. A Valedictory Address. Journal of the Royal Statistical Society 99, 674–713

Greville, T.N.E. (1943): Short Methods of Constructing Abridged Life Tables. Record of the American Institute of Actuaries 32(1), 29–42

Gueri, M., Gonzalea, C., Morin, V. (1986): The Effect of the Floods Caused by 'El Nino' on Health. Disasters 10, 118–124

Hafner, H., Haas, S., Pfeifer-Kurda, M., Eichhorn, S., Michitsuji, S. (1987): Abnormal Seasonality of Schizophrenic Births. A Specific Finding?. European Archives of Psychiatry and Clinical Neuroscience 236, 333–342

Hare, E.H. (1975): Season of Birth in Schizophrenia and Neurosis. American Journal of Psychiatry 132, 1168–1171

Hare, E., Moran, P. (1981): A Relation Between Seasonal Temperature and the Birth Rate of Schizophrenic Patients. Acta Physiologica Scandinavica 63, 396–405

Harris, B., (2001): Commentary: 'The Child is Father of the Man.' The Relationship Between Child Health and Adult Mortality in the 19th and 20th Century. International Journal of Epidemiology 30, 688–696

Haslam, N., Mayberry, J.F., Hawthorne, A.B., Newcombe, R.G., Holmes, G.K.T., Probert, C.S.J. (2000): Measles, Month of Birth and Crohn's Disease. Gut 47, 801–803

Hellmuth, K., Wnoroski, V. (1923): Variationsstatistischer Beitrag zur Frage des Einflusses der Jahreszeit auf das Köpergewicht der Neugeborenen, zugleich ein Hinweis auf die Bedeutung der Variationsstatistik bei der kritischen Bewertung von Sammelstatistiken in der Medizin. Klinische Wochenschrift 2, 75–78

Helgasson, T., Jonasson, M.R. (1981): Evidence for a Food Additive as a Case of Ketosis-prone Diabetes. Lancet ii, 716–720

Helsen, W.F., Starkes, J.L., van Winckel, J. (1998): The Influence of Relative Age on Success and Dropout in Male Soccer Player. American Journal of Human Biology 10(6), 791–798

Henderson, A.S., Korten, A.E., Jorm, A.F., McCusker, E., Creasy, H., Broe, G.A. (1991): Season of Birth in Alzheimer′s Disease in the Southern Hemisphere. Psychological Medicine 21, 371–374

Henderson, J.B., Dunnigan, M.G., McIntosh, W.B., Abdul-Motaal, A.A., Gettinby, G., Glekin, B.M. (1987): The Importance of Limited Exposure to Ultraviolet Radiation and Dietary Factors in the Aetiology of Asian Rickets: a Risk Factor Model. Quarterly Journal of Medicine 63, 413–425.

Higgins, C.D., Dos-Santos-Silva, I., Stiller, C.A., Swerdlow A.J. (2001): Season of Birth and Diagnosis of Children with Leukaemia. An Analysis of over 15 000 UK Cases occurring from 1953–95. British Journal of Cancer. 84(3), 406–412

Hillman, R.W., Conway, H.C. (1972): Season of Birth and Relative Body Weight. American Journal of Clinical Nutrition 25, 279–281

Hillman, R.W., Slaterand, P., Nelson, M.J. (1970): Season of Birth, Parental Age and Body Form. Some Interrelationships in Young Women. Human Biology 42, 570

Horiuchi, S. (1983): The Long-term Impact of War on Mortality. Old-age Mortality of the First World War Survivors in the Federal Republic of Germany. Population Bulletin of the United Nations 15, 80

Hu, Y.H., Kurioshi, T., Matsushita, Y., Nagata, C., Shimizu, H. (1996): Birth Season and Breast Cancer Risk in Japan. Breast Cancer Research and Treatment 39(3), 315–319

Hummel, M., Ziegler, A.G., Lewy, H., Ashkenazi, I., Laron, Z. (2001): IAA/GAD-positive Offspring of Diabetic Parents Have a Different Seasonality in Month of Birth Than Antibody-negative Offspring.(Letter) Diabetes Care 24(11)

Huntington, E. (1938): Season of Birth. J. Wiley & Sons, Inc., New York

Huxley R., Neil, A., Collins, R. (2002): Unravelling the Fetal Origins Hypothesis. Is There Really an Inverse Relationship Between Birthweight and Subsequent Blood Pressure?. Lancet 360(9334), 659–653

Ivarsson, A., Hernell, O., Nystrom, L., Persson, L.A. (2003): Children Born in the Summer Have Increased Risk for Coeliac Disease. Journal of Epidemiology and Community Health 57(1), 36–39

Jablonski, N., Chaplin, G. (2000) The Evolution of Human Skin Coloration. Journal of Human Evolution 39, 57–106.

James, W.H. (1978): Seasonality in Schizophrenia (letter), Lancet, 664

Jansson, B., Malahy, M.A. (1981): Cancer Risk, Age at Diagnosis, and Age at Death as Functions of Season of Birth. Cancer Detection and Prevention 4, 291–194

Joiner, T.E., Pfaff, J.J., Acres, J.G., Johnson, F. (2002) Birth Month and Suicidal and Depressive Symptoms in Australians Born in the Southern vs. the Northern Hemisphere. Psychiatric Research 112, 89–92.

Jongbloet, P.H. (1975): The Effects of Pre-ovulatory Overripness of Human Eggs on Development. In: Blandau, R.J. (Eds.): Aging Gametes. S. Karger. Basel, Switzerland, 300–329

Jongbloet, P.H., Greonewould, H.M.M., Hirasing, R.A., Vaan Buuren, S. (1998): Seasonality of Birth in Patients with Childhood Diabetes in the Netherlands. Diabetes Care 21, 190–191

Joseph, K.S., Kramer, M.S. (1996): Review of the Evidence of Fetal and Early Childhood Antecedents of Adult Chronic Disease. Epidemiologic Reviews 18(2), 158–174

Kannisto, V., Christensen, K., Vaupel, J.W. (1997): No Increased Mortality in Later Life for Cohorts Born During Famine. American Journal of Epidemiology 145(11), 987–994

Karl, T.R., Koss, W.J. (1984): Regional and National Monthly, Seasonal, and Annual Temperature Weighted by Area, 1895–1983. Historical Climatology Series 4–3, National Climatic Data Center, Asheville, NC

Keller, C.A., Nugent, R.P. (1983): Seasonal Patterns in Perinatal Mortality and Preterm Delivery. American Journal of Epidemiology 118 (5), 689–698

Kermack, W.O., McKendrick, A.G., McKinlay, P.L. (1934): Death-rates in Great Britain and Sweden. Some General Regularities and Their Significance. Lancet 226, 698–703

Kermack, W.O., McKendrick, A.G., McKinlay, P.L. (2001): Death–rates in Great Britain and Sweden. Some General Regularities and Their Significance. International Journal of Epidemiology 30, 678–683

Kettl, P.A., Collins, T., Sredy, M., Bixle, E.O. (1997): Seasonal Differences in Suicide Birth Rates in Alaska Natives Compared to Other Populations. American Indian and Alaska Native Mental Health Research 8, 1–10

Kida, K., Mimura, G., Ito, T., Murakami, K., Ashkenazi, I., Laron, Z. (2000): Incidence of Type 1 Diabetes (IDDM) in Children (0–14 Years) in Japan (1986–1990) Including an Analysis for Seasonality of Onset and Month of Birth. Diabetic Medicine 17, 59–63

Kinabo, J. (1993): Seasonal Variation of Birthweight Distribution in Morogoro, Tanzania. East African Medical Journal 70(12), 752–755

Kinlen, L.J., Willows, A.N. (1987): Cancer of the Testis and Month of Birth. British Journal of Cancer 55, 579–581

Kitagawa, E.M., Hauser, P.M. (1973): Differential Mortality in the United States. A Study in Socioeconomic Epidemiology. Harvard University Press, Cambridge, MA

Kliewer, E. (1992): Epidemiology of Diseases Among Migrants. International Migration, 141–64

Knox, A.G., Cummins, C. (1985): Birth Dates of Men with Cancer of the Testis. Journal of Epidemiology and Community Health 39, 237–243

Köppen, W. (1936): Das Geographische System der Klimate. In: Köppen, W., Geiger, R.: Handbuch der Klimatologie. Band 1, Teil C., Gebrüder Borntrager, Berlin

Kordonouri, O., Shuga, N., Lewy, H., Ashkenazi, I., Laron, Z. (2002): Seasonality of Month of Birth of Children and Adolescents with Type 1 Diabetes Mellitus in Berlin Differs from the General Population. European Journal of Pediatrics 161, 291–292

Korlath, J.A., Osterholmm, M.T., Judy, L.A., Forfang, J.C., Robinson, R.A. (1985): A Point-source Outbreak of Campylobacteriosis Associated with Consumption of Raw Milk. Journal of Infectious Diseases 152, 592–596

Koziel, S., Jankowska, E.A. (2002): Birtweight and Stature, Body Mass Index and Fat Distribution of 14-year-old Polish Adolescents. Journal of Paediatrics and Child Health 38(1), 55–58

Kramer, M.S. (2000): Invited Commentary: Association Between Restricted Fetal Growth and Adult Chronic Disease. Is it Causal? Is it Important? American Journal of Epidemiology 152(7): 605–608

Kramer, M.S. (1987): Determinants of Low Birthweight. Methodological Assessment and Meta-analysis. Bulletin of the World Health Organization 65, 663–737

Kristoffersen, S., Hartveit, F. (2000) Is a Woman's Date of Birth Related to her Risk of Developing Breast Cancer? Oncology Reports 7, 145–247

Kuh, D., Davey Smith, G. (1993): When is Mortality Risk Determined? Historical Insights Into a Current Debate. Social History of Medicine 6(1), 101–123

Kunugi, H., Watanabe, H., Sekiba, K., Kazamatsuri, H. (1998): Season of Birth of Chronic Alcoholics. Journal of Psychiatric Research 32, 321–323

Lam, D.A., Miron, J.A. (1996): The Effects of Temperature on Human Fertility. Demography 33, 291–306

Langlois, B., Blouin, C. (1998): Effect of a Horse´s Month Birth on its Future Sport Performance. II. Effect on Annual Earnings per Start. Annales de Zootechnie 47(1), 67–74

Laron, Z., Shamis, I., Nitzan-Kaluski, D., Ashkenazi, I. (1999): Month of Birth and Subsequent Development of Type 1 Diabetes (IDDM). Journal of Pediatric Endocrinology & Metabolism 12(3), 397–402

Lau, Y.I., Karlberg, J. (1998): Prevalence and Risk Factors of Childhood Asthma, Rhinitis and Eczema in Hong Kong. Journal of Paediatrics & Child Health 34(1), 47–52

Lechner, I., Mielck, A. (1998): Die Verkleinerung des "Healthy Migrant Effects": Entwicklung der Morbidität von ausländischen und deutschen Befragten im Sozio-Ökonomischen Panel 1984–1992. Gesundheitswesen 60, 715–20

Leon, D.A., Davey Smith, G. (2000): Infant Mortality, Stomach Cancer, Stroke, and Coronary Heart Disease. Ecological Analysis. British Medical Journal 320(7251), 1705–1706

Leon, D.A., Johansson, M., Rasmussen, F. (2000): Gestational Age and Growth Rate of Fetal Mass Are Inversely Associated with Systolic Blood Pressure in Young Adults. An Epidemiologic Study of 165,136 Swedish Men Aged 18 Years. American Journal of Epidemiology 152, 597–604

Lester, D. (1987): Month of Birth of Suicides, Homicides and Natural Deaths. Psychological Reports 60, 1310

Lester, D., Reeve, C.L., Priebe, K. (1970): Completed Suicide and Month of Birth. Psychological Reports 27, 210

Levenstein, H. (1988): Revolution at the Table. The Transformation of the American Diet. University Press, New York, Oxford

Levenstein, H. (1993): Paradox of Plenty. A Social History of Eating in Modern America. University Press, New York. Oxford

Levine, M.E., Wojcik, B.E. (1999): Alcoholic Typology and Season of Birth. Journal of Addictive Disorders 18, 41–52

Lewis, M.S., Griffin, P.A. (1981): An Explanation for the Season of Birth Effect in Schizophrenia and Other Diseases. Psychological Bulletin 89, 589–596

Lillard, L.A., Panis, C.W.A. (1996): Marital Status and Mortality. The Role of Health (in Mortality). Demography 33(3), 313–327

Liang, K.–Y., Zeger, S.L. (1986): "Longitudinal Data Analysis Using Generalized Linear Models", Biometrika, 13–22

London, W.P. (1998): Alcoholism: Theoretical Consideration of Season of Birth and Geographic Latitude. Alcohol 4, 127–129

Lucas, A., Fewtrell, M.S., Cole, T.J. (1999): Fetal Origins of Adult Disease. The Hypothesis Revisited. British Medical Journal 319, 245–249

Lummaa, V., Lemmetyinen, R., Haukioja, E., Pikkola, M. (1998): Seasonality of Births in Homo Sapiens in Preindustrial Finland. Maximisation of Offspring Survivorship?. Journal of Evolutionary Biology 11, 147–157

Madico, G., McDonald, J., Gilman, R.H., Cabrera, L., Sterling, C.R. (1997): Epidemiology and Treatment of Cyclospora Cayetanenis Infection in Perucian Children. Clinical Infectious Diseases 24, 977–981

Maheswaran, R., Strachan, D.P., Dodgeon, B., Best, G.N. (2002): A Population-based Case-control Study for Examining Early Life Influences on Geographical Variation in Adult Mortality in England and Wales Using Stomach Cancer and Stroke as Examples. International Journal of Epidemiology 31(2), 375–382

Mason, W.M., Smith, H.L. (1985): Age-period-cohort Analysis and the Study of Deaths from Pulmonary Tuberculosis. In: Mason, W.M., Fienberg S.E., (Eds): Cohort Analysis in Social Research. Beyond the Identification Problem. Springer-Verlag, New York, 151–227

Matsuada, S., Sone, T., Doi, T., Kahyo, H. (1993): Seasonality of Mean Birth-weight and Mean Gestational Period in Japan. Human Biology 65(3), 481–501

Mattock, C., Marmot, M., Stern, G. (1988): Could Parkinson´s Disease Follow Intra-uterine Influenza? A Speculative Hypothesis. Journal of Neurology, Neurosurgery and Psychiatry 35, 753–756

McGrath, J., Welham, J.L. (1999): Season of Birth and Schizophrenia: a Systematic Review and Meat-analysis of Data from the Southern Hemisphere. Schizophrenia Research 35, 237–242

McGrath, J., Selten, J.P., Chan, D. (2002): Long-term Trends in Sunshine and its Association with Schizophrenia Birth Rates and Age at First Registration –

Data from Australia and the Netherlands. Schizophrenia Research 54(3), 199–212

McGrath, J., Eyles, D., Mowry, B., Yolken, R., Buka, S. (2003): Low Maternal Vitamin D as a Risk Factor for Schizophrenia: a Pilot Study Using Banked Sera. Schizophrenia Research 63(1–2), 73–78

McNeil, T., Dalen, P., Dzierzykray-Rogalska, M., Kaij, L. (1975): Birth Rates of Schizophrenics Following Relatively Warm Versus Relatively Cool Summers. Archiv für Psychiatrie und Nervenkrankheiten 221, 1–10

Mills, C.A. (1941): Mental and Physical Development as Influenced by Season of Conception. Human Biology 13, 378–389

Miura, T. (1987): Seasonality of Birth. SPB Academic Publishing, The Hague

Miura, T., Shimura, M., Kimura, T. (1987): Season of Birth in Parkinsonism. In: Miura, T. (Eds): Seasonality of Birth. Progress in Biometeorology. Hague, Netherlands, 157–162

Modestin, J., Ammann, R., Wurmle, O. (1995): Season of Birth. Comparison of Patients with Schizophrenia, Affective Disorders, and Alcoholism. Acta Psychatrica Scandinavia 91, 140–143

Moore, S.E. (1998): Nutrition, Immunity and the Fetal and Infant Origins of Disease Hypothesis in Developing Countries. Proceedings of the Nutrition Society 57, 241–247

Moore, S.E., Cole, T.J., Poskitt, E.M.E., Sonko, B.J., Whitehead, R.G., McGregor, I.A., Prentice, A.M. (1997): Season of Birth Predicts Mortality in Rural Gambia. Nature 338, 434

Morgan, R. (1978): Seasonality in Schizophrenia (letter). Lancet, 830

Nakao, H. (1988): Season of Birth Breast Cancer Patients and its Relation to Patient's Reproductive History in Tokyo, Japan. Acta Medica Okayama 42(4), 231–241

Nelson, W., Tong, Y.L., Lee, J.-K., Halberg, F. (1979): Methods for Cosinor Rhythmometry. Chronobiologia 6, 305–23

Neu, A., Kehrer, M., Ashkenazi, I., Laron, Z. (2000): Seasonality of Birth in Children (0–14 Years) with Diabetes Mellitus Type 1 in Baden-Württemberg, Germany. The Journal of Pediatric Endocrinology & Metabolism 13, 1081–1085

Nielsen, S. (1992): Seasonal Variation in Anorexia Nervosa? Some Preliminary Findings from a Neglected Area of Research. The International Journal of Eating Disorders 11, 25–35

Nielsen, L.L., Nielsen, N.M., Melbye, M., Sodermann, M., Jacobsen, M., Aaby, P. (1998): Exposure to Measles in Utero and Crohn's Disease. Danish Register Study. British Medical Journal 316, 196–197

Nilsson, L., Bjorksten, B., Hattevig, G., Kjellman, B., Sigurs, N., Kjellman, N.I.M. (1997): Season of Birth as Predictor of Atopic Manifestations. Archives of Disease in Childhood 76(4), 341–344

Nonaka, K., Imaizumi, Y. (2000): Deaths from Cerebrovascular Diseases Correlated to Month of Birth. Elevated Risk of Death from Subarachnoid Hemorrhage Among Summer Born. International Journal of Biometeorology 44, 182–185

Norjavaara, E., de Verdier, M.G., Lindmark, B. (2001): Adult Height in Women with Childhood Asthma – a Population-based Study. Pharmacoepidemiology and Drug Safety 10 (2), 121–125

Orford, S., Dorling, D., Mitchell, R. et al. (2002): Life and Death of the People of London. A Historical GIS of Charles Booth's Inquiry. Health Place 8(1), 25–35

Osmond, C., Barker, D.J.P., Slattery, J.M. (1990): Risk of Death from Cardiovascular Disease and Chronic Bronchitis Determined by the Place of Birth in England and Wales. Journal of Epidemiology and Community Health 44(2), 139–141

Otto, W., Glaass, R., (1959): Inwieweit sind Tragezeit, Körperlänge und Körpergewicht der Neugeborenen von jahreszeitlichen Einflüssen abhängig? Deutsches Gesundheitswesen 14(35), 1625–1628

Otto, W., Noack, H. (1957): Jahreszeitliche Schwankungen der Körpermasse bei 100000 neugeborenen Berliner und Leipziger Kindern. Zur menschlichen Vererbungs- und Konstitutionslehre 34, 250–260

Pallast, E.G.M., Jongbloet, P.H., Straatman, H.M., Zielhuis, G.A. (1994): Excess Seasonality of Births Among Patients with Schizophrenia and Seasonal Ovopathy. Schizophrenia Bulletin. 129, 355–361

Pasamanick, B. (1986): Seasonality of Schizophrenia Births (letter). American Journal of Orthopsychiatry 56, 168–169

Penas-Lledo, E., Waller, G. (2002): Pattern of Birth and Eating Attitudes in Young Adults. Failure to Replicate in a Warmer Climate. International Journal of Eating Disorders 32, 367–71

Pflug, E.J.S. (2001): Season of Birth, Schooling and Earnings. Journal of Economic Psychology 22(5), 641–660

Phillips, D.I.W. (1993): Twin-studies in Medical Research – Can They Tell us whether Diseases are Genetically Determined. Lancet 341, 1008–1009

Philips, D.I.W., Young, J.B. (2000): Birthweight, Climate at Birth and the Risk of Obesity in Adult Life. International Journal of Obesity 24, 281–287

Philpot, M., Rottenstein, M., Burns, A. (1989): Season of Birth in Alzheimer's Disease. The British Journal of Psychiatry 155, 662–666

Pokorny, A.D. (1960): Characteristics of 44 Patients who Subsequently Committed Suicide. Archives of General Psychiatry 2, 314–323

Pollitt, E.A.J (1989): Seasonality and Weight Gain During the First Year of Life. American Journal of Human Biology 1, 747–756

Prener, A., Carstensen, B. (1990): Month of Birth and Testicular Cancer Risk in Denmark. American Journal of Epidemiology 131(1), 15–19

Prentice, A.M., Whitehead, R.G., Roberst, S.B., Paul, A.A. (1981): Long Term Energy Balance in Childbearing Gambian Women. American Journal of Clinical Nutrition 34, 2790–2799

Preston, S.H., Haines, M.R. (1991): Fatal Years. Child Mortality in Late Nineteenth-century in America. Princton University Press, Princton

Preston, S.H., Hill, M.E., Drevenstedt, G.L. (1998): Childhood Conditions That Predict Survival to Advanced Ages Among African Americans. Social Science & Medicine 47 (9), 1231–1246

Preston, S.H., van de Walle, E. (1978): Urban French Mortality in the Nineteenth Century. Population Studies 32(2), 275–297

Pringle, D.G. (1998): Hypothesized Foetal and Early Life Influences on Adult Heart Disease Mortality. An Ecological Analysis of Data for the Republic of Ireland. Social Science & Medicine 46(6), 683–693

Procopio, M., Marriott, P.K. (1998): Seasonality of Birth in Epilepsy. A Danish Study. Acta Neurologica Scandinavica 98, 297–301

Quested, D.J. (1991): Incidence Rates of Schizophrenia (letter). British Journal of Psychiatry 159, 587–588

Razum, O., Zeeb, H., Akgün, H.S., Yilmaz, S. (1998): Low Mortality of Turkish Residents in Germany Persists and Extends Into the Second Generation. Merely a Healthy Migrant Effect?. Tropical Medicine & International Health 3, 297–303

Regidor, E., Gutierrez-Fisac, J.L., Calle, M.E. et al. (2002): Infant Mortality at Time of Birth and Cause-specific Adult Mortality Among Residents of the Region of Madrid Born. International Journal of Epidemiology 31 (2), 368–374

Reinsch, H. (1967): Smoothing by Spline Functions, Numerische Mathematik 10, 177–83

Rezaul, I., Persaud, R., Takei, N., Treasure, J. (1996): Season of Birth and Eating Disorders. The International Journal of Eating Disorders 19, 53–61

Roberts, S., Paul, A.A., Cole, T.J., Whitehead, R.G. (1982): Seasonal Changes in Activity, Birthweight, and Lactational Performance in Rural Gambian Women. Transactions of the Royal Society of Tropical Medicine and Hygiene 76, 668–678

Roseboom, T.J., van der Meulen, J.H.P., Osmond, C., Barker, D.J., Ravelli, A.C., Bleker, O.P. (2000a): Plasma Lipid Profiles in Adults After Prenatal Exposure to the Dutch Famine. American Journal of Clinical Nutrition 72, 1101–1106

Roseboom, T.J., van der Meulen, J.H.P., Osmond, C., Barker, D.J., Ravelli, A.C., Schroeder-Tanka, J.M. et al. (2000b): Coronary Heart Disease After Prenatal Exposure to the Dutch Famine, 1944–45. Heart 84, 595–598

Roseboom, T.J., van der Meulen, J.H.P., Osmond, C. et al. (2001): Adult Survival After Prenatal Exposure to the Dutch Famine 1944–45. Paediatric and Perinatal Epidemiology

Rothwell, P.M., Gutnikow, S.A., McKinney, P.A., Schober, E., Ionescu-Rousham, E.K., Gracey, M. (1998): Seasonality of Low Birthweight in Indigenous Australians. An Increase in Pre-term Birth or Intrauterine Growth Retardation?. Australian & New Zealand Journal of Public Health 22(6), 669–672

Sadovnick, A.D., Yee, I.M.L. (1994): Season of Birth in Multiple Sclerosis. Acta Neurologica Scandinavia 89, 190–191

Salemi, G., Ragonese, P., Aridon, P., Reggio, A., Nicoletti, A., Buffa, D., Conte, S., Savettieri, G. (2000): Is Season of Birth Associated with Multiple Sclerosis?. Acta Neurologica Scandinavica 101(6), 381–383

Salib, E. (2002): Month of Birth and Suicide. An Exploratory Study. International Journal of Psychiatry in Clinical Practice 6(1), 39–44

Samuelsson, U., Johansson, C. Ludvigsson, J. (1999): Month of Birth and Risk of Developing Insulin Dependent Diabetes in Southeast Sweden. Archives of Disease of Childhood 81, 143–146

Sanborn, D.E., Sanborn, C.J. (1974): Suicide and Month of Birth. Psychological Reports 34, 950

Schäfer, T., Przybilla, B., Ring, J., Kunz, B., Greif, A., Ueberla, K. (1993): Manifestation of Atopy is not Related to Patient's Month of Birth. Allergy 48(4), 291–294

Schneider, D., Greenberg, M., Lu, L.L. (1997a): Early Life Experiences Linked to Diabetes Mellitus. A Study of African-American Migration. Journal of the National Medical Association 89(1), 29–34

Schneider, D., Greenberg, M.R., Lu, L.L. (1997b): Region of Birth and Mortality from Circulatory Diseases Among Black Americans. American Journal of Public Health 87(5), 800–804

Schwarz, F., Spielauer, M., Städtner, K. (2002a): The Effect of Gender, Residential Area, Education of Parents and Previous Schooling on University Education. ÖIF Working Paper 27–2002, Austrian Institute for Family Studies, Vienna

Schwarz, F., Spielauer, M., Städtner, K. (2002b): Gender, Regional and Social Differences at the Transition from Lower to Upper Secondary Education. ÖIF Working Paper 23–2002, Austrian Institute for Family Studies, Vienna

Seiver, D.A. (1985) Trend and Variation in the Seasonality of U.S. Fertility, 1947–1976. Demography 22(1), 89–100

Shears, P., Wright, A. (1995): Community-acquired Infections Among Children in an Urban Environment. A 2-year Prospective Study in Liverpool, UK. Journal of Infection 30(2), 173–177

Shimura, M., Kimura, T., Miura, T. (1987): Season of Birth in Some Neurological Disorders – Multiple Sclerosis, Amyotrophic Lateral Sclerosis, Senile Dementia. In: Miura, T. (Eds.): Seasonality of Birth. Progress in Biometeorology. SPÄB Academic Publishing, The Hague, The Netherlands

Smithers, A.G., Cooper, H.J. (1984): Social Class and Season of Birth. Journal of Social Psychology 124, 79–84

Songini, M., Casu, A., Ashkenazi, I., Laron, Z. (2001): Seasonality of Birth in Children (0–14 Years) and Young Adults (0–29 Years) with Type 1 Diabetes Mellitus in Sardinia Differs from That in the General Population. The Journal of Pediatric Endocrinology & Metabolism 14(6), 781–783

Sonnenberg, A., Mueller, H., Pace, F. (1985): Birth-cohort Analysis of Peptic Ulcer Mortality in Europe. Journal of Chronic Diseases 38, 309–317

Sonnenberg, A. (1987): Causative Factors in the Etiology of Peptic Ulcer Disease Become Effective Before the Age of 15 Years. Journal of Chronic Disease 40, 193–202

Sorensen, H.T., Pedersen, L., Olsen, J.H., Rothman, K.J. (2001): Seasonal Variation in Month of Birth and Diagnosis of Early Childhood Acute Lymphoblastic Leukemia. Journal of the American Medical Association 285(2), 168–169

Speare, A., Meyer, J.W. (1988): Types of Elderly Residential Mobility and Their Determinants. Journal of Gerontology Social Sciences 43, S71–S74

Sparén, P., Vågerö, D., Shestov, D.B., Plavinskaja, S., Parfenova, N., Hoptiar, V., Paturot, D., Galanti, M.R. (2003): Long Term Mortality After Severe Starvation During the Siege of Leningrad: Prospective Cohort Study. British Medical Journal, doi:10.1136/bmj.37942.603970.9a(published 5 December 2003)
Stanner, S.A., Bulmer, K., Andres, C., Lantseva, O.E., Borodina, V., Poteen, V.V. et al. (1997): Does Malnutrition In Utero Determine Diabetes and Coronary Heart Disease in Adulthood? Results from the Leningrad Siege Study, a Cross Sectional Study. British Medical Journal 315, 1342–1348
Strachan, D.P., Leon, D.A., Dodgeon, B. (1995): Mortality from Cardiovascular Disease Among Interregional Migrants in England and Wales. British Medical Journal 310(6977), 423–427
Strachan, D.P., Seagroatt, V., Cook, D.G. (1994): Chest Illness in Infancy and Chronic Respiratory Disease in Later Life. An Analysis by Month of Birth. International Journal of Epidemiology 23(5), 1060–1068
Susser, M., Stein, Z. (1962): Civilication and Peptic Ulcer. Lancet 1:115–119
Susser, M. (2001): The Longitudinal Perspective and Cohort Analysis. International Journal of Epidemiology 30, 684–687
Swerdlow, A.J. (1991): Mortality and Cancer Incidence in Vietnamese Refugees in England and Wales. A Follow-up Study. International Journal of Epidemiology 20, 13–19
Templer, D.I., Trent, N.H., Spencer, D.A., Trent, A., Corgiat, M.D., Mortensen, P.B., Gorton, M. (1992): Season of Birth in Multiple Sclerosis. Acta Neurologica Scandinavica 85(2), 107–109
Thors, H., Svanes, C., Thjodleifsson, B. (2002): Trends in Peptic Ulcer Morbidity and Mortality in Iceland. Journal of Clinical Epidemiology 55(7), 681–686
Ticher, A., Ring, A., Barak, Y., Elizur, A., Weizman, A. (1996): Circannual Pattern of Autistic Births. Reanalysis in Three Ethnic Groups. Human Biology 68, 585–592
Tirgoviste, C. (1999): Seasonality of Birth in Children with Diabetes in Europe. Multicentre Cohort Study. British Medical Journal 319, 887–888
Toft-Sorensen, H., Pedersen, L., Norgard, B., Fonager, K., Rothmann, K.J. (2001): Does Month of Birth Affect Risk of Crohn's Disease in Childhood and Adolescence?. British Medical Journal 323, 907
Torrey, E.F., Bowler, A.F., Rawlings, R. (1991): An Influenza Epidemic and the Seasonality of Schizophrenic Births. In: Kurstak, E. (Eds): Psychiatry and Biological Factors. Plenum Medical Book Company, New York, 109–116
Torrey, E.F., Miller, J., Rawlings, R., Yolken, R.H. (1997): Seasonality of Births in Schizophrenia and Bipolar Disorders: A Review of the Literature. Schizophrenia Research 28, 1–38
Torrey, E.F., Torrey, B.B., Peterson, M.R. (1977): Seasonality of Schizophrenic Births in the United States. Archives of General Psychiatry 34, 1065–1070
Townbridge, F.L., Newton, L.H. (1979): Seasonal Changes in Malnutrition and Diarrheal Disease Among Preschool Children in El Salvador. American Journal of Tropical Medicine and Hygiene 28, 135–141
Turnquist, K. (1993): Second-trimester Markers of Fetal Size in Schizophrenian (letter). American Journal of Psychiatry 150. 1571–1572

Ursic-Bratina, N., Battelino, T., Krzisnik C., Laron-Kenet, T., Ashkenazi, I., Laron, Z. (2001): Seasonality of Birth in Children (0–14 Years) with Diabetes Mellitus Type 1 in Slovenia. Journal of Pediatric Endocrinology & Metabolism 14, 47–52

Vaiserman, A.M., Collinson, A.C., Koshel, N.M., Belaja, I.I., Voitenko, V.P. (2002): Seasonal Programming of Adult Longevity in Ukraine. International Journal of Biometeorology 47, 49–52

Valkonen, T. (1989): Adult Mortality and Level of Education. A Comparison of Six Countries. In: Fox, J. (Eds.): Health Inequalities in European Countries. Aldershot: Gower Publishing, 142–162

Vaupel, J.W., Gambill, B.A., Yashin, A.I. (1987): Thousands of Data at a Glance – Shaded Contour Maps of Demographic Surfaces. Population Index 53(3), 376–376

Vaupel, J.W., Wang, Z., Andreev, K.F., Yashin, A.I. (1997): Population Data at a Glance. Shaded Contour Maps of Demographic Surfaces Over Age and Time. Odense Monographs on Population Aging, 4, Odense University Press, Odense

Vaupel, J.W., Yashin, A.I. (1985): Heterogeneity's Ruses. Some Surprising Effects of Selection in Population Dynamics. The American Statistician 39, 176–185

Vezina, H., Houde, L., Charbonneau, H., Beaudry, M., Cholette, A., Daoud, N., Mathieu, J., Robitaille, Y., Veilleux, F., Gauvreau, D. (1996): Season of Birth and Alzheimer's Disease. A Population-based Study in Saguenay-Lac-St-Jean, Quebec (IMAGE-Project). Psychological Medicine 26(1), 143–149

Vitiello, B., Hill, J.L., Molchan, S.E., Martinez, R.A., Martinson, H.J., Sunderland, T. (1991): Lack of Seasonal Variation in the Birth of Patients with Dementia of the Alzheimer's Type. Psychiatry Research 39, 21–24

Waaler, H.F. (1984): Height, Weight and Mortality – The Norwegian Experience. Acta Neurologica Scandinavia, Suppl. 679

Waller, G., Watkins, B., Potterton, C., Niederman, M., Sellings, J., Willoughby, K. et al. (2002): Pattern of Birth in Adults with Anorexia Nervosa. Journal of Nervous and Mental Diseases 190, 752–756

Ward, W.P. (1987): Weight at Birth in Vienna, Austria, 1865–1930. Annals of Human Biology 14(6), 495–506

Watkins, B., Willoughby, K., Aller, G., Serpell, L., Lask, B. (2002): Pattern of Birth in Anorexia Nervosa I. Early-onset Cases in the United Kingdom. International Journal of Eating Disorders 32, 11–17

Watson, C.G., Kucala, T., Tilleskjor, C., Jacobs, L. (1984): The Birth Effect in Nonschizophrenic Psychiatric Patients. Journal of Clinical Psychology 40, 884–888

Watson, C.G., Kucala, T., Tilleskjor, C., Jacobs, L. (1984): Schizophrenic Birth Seasonality in Relation to the Incidence of Infectious Diseases and Temperature Extremes. Archives of General Psychiatry 41, 85–90

Weber, G.H., Prossinger, H., Seidler, H. (1998): Height Depends on Month of Birth. Nature 391, 754–755

Wehr, T. (1998): Effect of Seasonal Changes in Daylength on Human Neuroendocrine Function. Hormone Research 49, 118–124

Wendl-Richter, H.U. (1997): Birthweight Distribution in Rural North-west Burkina Faso. Tropical Medicine & International Health 2(4), 404–408

Weitoft, G., Gullberg, A., Hjern, A., Rosén, M. (1999): Mortality Statistics in Immigrant Research: Method for Adjusting Underestimation of Mortality. International Journal of Epidemiology 28(4), 756–763

Wiberg, M., Templer, D.I. (1994): Season of Birth in Multiple Sclerosis in Sweden. Replication. Journal of Orthomolecular Medicine 9(2), 71–74

Willoughby, K., Watkins, B., Beumont, P., Maguire, S., Lask, B., Waller, G. (2002): Pattern of Birth in Anorexia Nervosa II. A Comparison of Early-onset Cases in the Southern and Northern Hemispheres. International Journal of Eating Disorders 32, 18–23

Wilmoth, J.R. (1986): Age-Period-Cohort Analysis of Mortality Data. An Exploratory Approach. Population Index 52(3), 416–416

Wilmoth, J., Vallin, J., Caselli, G. (1989): Differences Between Observed and Expected Cohort Mortality. Population 44(2), 335–376

Willis, J.A., Scott, R.S., Darlow, B.A., Lewy, H., Ashkenazi, I., Laron, Z. (2002): Seasonality of Birth and Onset of Clinical Disease in Children and Adolescents (0–19 Years) with Type 1 Diabetes Mellitus in Canterbury, New Zealand. Journal of Pediatric Endocrinology & Metabolism 15(5), 645–647

Wohlfahrt, J. et al. (1998): Secular and Seasonal Variation of Length and Weight at Birth. Lancet 352, 1990

Xu, X. et al. (2001): Seasonality of Growth in Shanghai Infants (n=4128) Born in 11 Consecutive Years. European Journal of Clinical Nutrition 55(8), 714–725

Ye, J., Chen, R.-G., Ashkenazi, I., Laron, Z. (1998): Lack of Seasonality in Childhood IDDM (0,7 –15 Years) in Shanghai, China. The Journal of Pediatric Endocrinology & Metabolism 11, 461–464

Young, C.M. (1987): Migration and Mortality. The Experience of Birthplace Groups in Australia. International Migration Review 21(3), 531–554

Yuen, J., Ekbom, A., Trichopoulos, D., Hsieh, C.-C., Adami, H.-O. (1994): Season of Birth and Breast Cancer Risk in Sweden. British Journal of Psychiatry 70, 564–568

Printing: Strauss GmbH, Mörlenbach
Binding: Schäffer, Grünstadt